Solutions Manual to Accompany

Modern Engineering Statistics

Solutions Manual to Accompany

Modern Engineering Statistics

Thomas P. Ryan
Acworth, GA

WILEY-INTERSCIENCE
A JOHN WILEY & SONS, INC., PUBLICATION

Library of Congress Cataloging in Publication Data:

Ryan, Thomas P.
    Solutions Manual to Accompany Modern Engineering Statistics

    ISBN 978-0-470-09607-9 (paper)

Contents

Methods of Collecting and Presenting Data

Note: The data in the following exercises, including data in MINITAB files (i.e., the files with the .MTW extension), can be found at the website for the text: ftp://ftp.wiley.com/public/sci_tech_med/engineering_statistics. This also applies to the other chapters in the text.

1.1. Given below are the earned run averages (ERAs) for the American League for 1901-2003 (in ERAMC.MTW), with the years 1916 and 1994 corrected from the source, *Total Baseball*, 8th edition, by John Thorn, Phil Birnbaum, and Bill Deane, since those two years were obviously in error. (The league started in 1901.)

```
Year   1901   1902   1903   1904   1905   1906   1907   1908
       1909   1910   1911   1912   1913   1914
ERA    3.66   3.57   2.96   2.60   2.65   2.69   2.54   2.39
       2.47   2.51   3.34   3.34   2.93   2.73
Year   1915   1916   1917   1918   1919   1920   1921   1922
       1923   1924   1925   1926   1927   1928
ERA    2.93   2.82   2.66   2.77   3.22   3.79   4.28   4.03
       3.98   4.23   4.39   4.02   4.14   4.04
Year   1929   1930   1931   1932   1933   1934   1935   1936
       1937   1938   1939   1940   1941   1942
ERA    4.24   4.64   4.38   4.48   4.28   4.50   4.45   5.04
       4.62   4.79   4.62   4.38   4.15   3.66
Year   1943   1944   1945   1946   1947   1948   1949   1950
       1951   1952   1953   1954   1955   1956
ERA    3.29   3.43   3.36   3.50   3.71   4.29   4.20   4.58
       4.12   3.67   3.99   3.72   3.96   4.16
Year   1957   1958   1959   1960   1961   1962   1963   1964
       1965   1966   1967   1968   1969   1970
ERA    3.79   3.77   3.86   3.87   4.02   3.97   3.63   3.62
       3.46   3.43   3.23   2.98   3.62   3.71
Year   1971   1972   1973   1974   1975   1976   1977   1978
       1979   1980   1981   1982   1983   1984
ERA    3.46   3.06   3.82   3.62   3.78   3.52   4.06   3.76
       4.21   4.03   3.66   4.07   4.06   3.99
Year   1985   1986   1987   1988   1989   1990   1991   1992
       1993   1994   1995   1996   1997   1998
ERA    4.15   4.17   4.46   3.96   3.88   3.90   4.09   3.94
       4.32   4.80   4.71   5.00   4.57   4.65
Year   1999   2000   2001   2002   2003
ERA    4.86   4.91   4.47   4.46   4.52
```

Construct a time sequence plot, either by hand or using software such as MINITAB, or equivalently a scatterplot with ERA plotted against Year. Does the plot reveal a random pattern about the overall average for these 103 years, or does the plot indicate nonrandomness and/or a change in the average?

Solution:
Here is the time sequence plot that is actually in the form of a scatterplot.

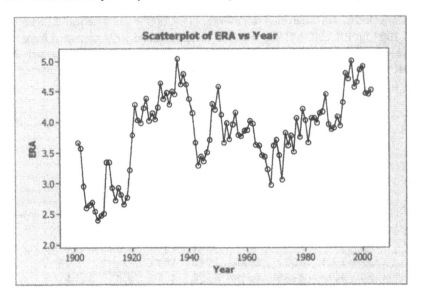

(a) There is considerable nonrandomness in this plot, especially the strong upward trend since about 1970 as well as the monotonicity during certain intervals of years (e.g., strictly decreasing from 1938-43.)

1.3. Construct a dotplot for the data in Example 1.2.
Solution:

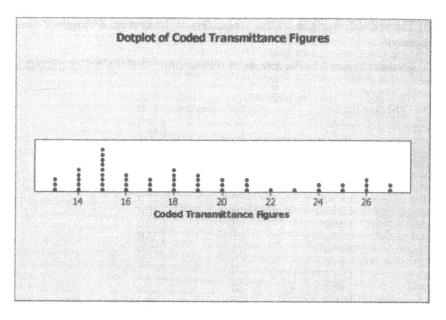

1.5. Statistical literacy is important not only in engineering but also simply as a means of expression. There are many statistical guffaws that appear in lay publications. Some of these are given in the "Forsooth" section of *RSS News* (Royal Statistical Society News) each month. Others can be found online at *Chance News*, whose website is at the following URL: http://www.dartmouth.edu/~chance/chance_news/news.html. The following two statements can be found at the latter website. Explain what is wrong with each statement.

(a) Migraines affect approximately 14% of women and 7% of men, that's one-fifth the population (*Herbal Health Newsletter Issue 1*)

(b) Researchers at Cambridge University have found that supplementing with vitamin C may help reduce the risk of death by as much as 50% (*Higher Nature Health News* No. HN601, 2001).

(*Comment*: Although the errors in these two statements should be obvious, misstatements involving statistical techniques are often made, even in statistics books, that are not obvious unless one has a solid grasp of statistics.)

Solution:
(a) The percentages are not additive since neither is based on 100% of the population of men and women combined. More specifically, even if the number of women was the same as the number of men, the base would be doubled if the populations were combined, rather than staying at the common number, as would be necessary for the percentages to be additive.

(b) The risk will always be 100% ... regardless of the amount of Vitamin C that is ingested!

1.7. Consider Figure 1.5. The data are as follows (in 75-25PERCENTILES2002.MTW):

Row	SAT 75th percentile	acceptance rate	SAT 25th percentile
1	1540	12	1350
2	1580	11	1410
3	1550	16	1380
4	1580	13	1450
5	1560	16	1410
6	1560	13	1360
7	1490	23	1310
8	1500	26	1300
9	1510	13	1310
10	1520	21	1330
11	1490	44	1280
12	1470	33	1290
13	1510	23	1310
14	1450	30	1290
15	1490	16	1290
16	1480	32	1300
17	1460	45	1300
18	1460	31	1270
19	1430	34	1270
20	1450	26	1200
21	1410	39	1200
22	1400	55	1220
23	1460	36	1280
24	1400	29	1170
25	1380	49	1220
26	1410	26	1240
27	1340	37	1130
28	1410	41	1230
29	1370	38	1160
30	1450	22	1280
31	1420	29	1250
32	1420	48	1220
33	1400	34	1210
34	1410	50	1240
35	1390	32	1220
36	1450	71	1240
37	1365	46	1183
38	1420	57	1250
39	1290	63	1060
40	1275	57	1070
41	1330	79	1130
42	1270	78	1050
43	1290	48	1080
44	1350	36	1150
45	1370	73	1180
46	1290	66	1070
47	1290	47	1090
48	1310	62	1090
49	1390	73	1210

(a) Construct the graph of the acceptance rate against the 75th percentile SAT score with the latter on the horizontal axis. Is the slope exactly the same as the slope of Figure 1.5? Explain why the slope should or should not be the same.

(b) Construct the graph of the 25th percentile SAT score against the acceptance rate with the former on the vertical axis. (The data on the 25th percentile are in the third column in the file.) Does the point that corresponds to point #22 in Figure 1.5 also stand out in this graph?

(c) Compute the difference between the 75th percentile and 25th percentile for each school and plot those differences against the acceptance rate. Note that there are two extreme points on the plot, with differences of 250 and 130, respectively. One of these schools is for a prominent public university and the other is a private university, both in the same state. Which would you guess to be the public university?

Solution:
(a) No, the slopes (of a line fit through the points, for example) will not be the same because the axes are reversed.

(b) No, there are no points on the graph that stand out as being unusual.

(c) We would guess that the point with the higher acceptance rate would be the public university (which it is: University of California --Berkeley)

1.9. Consider different amounts of one-dimensional data. What graphical display would you recommend for each of the following numbers of observations: (a) 10, (b) 100, and (c) 1000?

Solution:
(a) dotplot (b) dotplot or histogram (c) histogram

1.11. The following numbers are the first 50 of 102 chemical data measurements of color from a leading chemical company that were given in Ryan (2000): 0.67, 0.63, 0.76, 0.66, 0.69, 0.71, 0.72, 0.71, 0.72, 0.72, 0.83, 0.87, 0.76, 0.79, 0.74, 0.81, 0.76, 0.77, 0.68, 0.68, 0.74, 0.68, 0.68, 0.74, 0.68, 0.69, 0.75, 0.80, 0.81, 0.86, 0.86, 0.79, 0.78, 0.77, 0.77, 0.80, 0.76, 0.67, 0.73, 0.69, 0.73, 0.73, 0.74, 0.71, 0.65, 0.67, 0.68, 0.71, 0.69, and 0.73.

(a) What graphical display would you suggest if it was suspected that there may be some relationship between consecutive measurements (which would violate one of the assumptions of the statistical methods presented in later chapters)?

(b) Construct the display that you suggested in part (a). Do consecutive observations appear to be related?

Solution:

(a) Time sequence plot or autocorrelation plot, preferably the latter.

(b) The autocorrelation function plot (the correlation between units one unit apart, two units apart ...) is given below. There does appear to be autocorrelation, especially of observations one or two units apart since the autocorrelations for the first two lags are above the upper decision line (95% confidence interval) and the *t*-statistics exceed 2, as can be seen from the numbers below the graph.)

Autocorrelation Function for C1

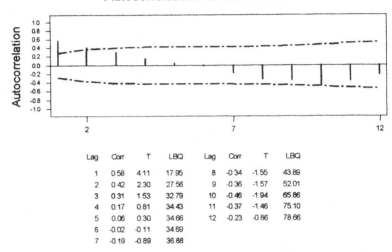

Lag	Corr	T	LBQ	Lag	Corr	T	LBQ
1	0.58	4.11	17.95	8	-0.34	-1.55	43.89
2	0.42	2.30	27.56	9	-0.36	-1.57	52.01
3	0.31	1.53	32.79	10	-0.46	-1.94	65.86
4	0.17	0.81	34.43	11	-0.37	-1.46	75.10
5	0.06	0.30	34.66	12	-0.23	-0.86	78.66
6	-0.02	-0.11	34.69				
7	-0.19	-0.89	36.88				

1.13. This exercise illustrates how the choice of the number of intervals greatly influences the shape of a histogram. Construct a histogram of the first 100 positive integers for the following numbers of classes: 3, 4, 6, 7, 10, and 12. (The number of classes can be specified in MINITAB, for example, by using the NINT subcommand with the HIST command, and the sequence MTB>SET C1, DATA>1:100, DATA>END will place the first 100 integers in the first column of the worksheet.) We know that the distribution of numbers is uniform over the integers 1-100 because we have one of each. We also have the same number of observations in the intervals 1-9, 10-19, 20-29, and so on. Therefore, the histograms should theoretically be perfectly flat. Are any of the histograms flat? In particular, what is the shape when only three classes are used? Explain why this shape results. What does this exercise tell you about relying on a histogram to draw inferences about the shape of the population of values from which the sample was obtained?

Solution:

(simulation exercise by student which shows that histograms are not reliable indicators for the shape of population distributions)

1.15. Explain why consecutive observations that are correlated will be apparent from a digidot plot but not from a dotplot, histogram, stem-and-leaf display, scatter plot, or boxplot. Is there another plot that you would recommend for detecting this type of correlation? Explain.

Solution:
There is no time order involved in a dotplot, histogram, stem-and-leaf display, scatter plot, or boxplot. A time sequence plot would be another possibility.

1.17. Construct a box plot of your driving times from the previous problem. Do any of your times show as an outlier? If the box doesn't exhibit approximate symmetry, try to provide an explanation for the asymmetry.

Solution:
(box plot to be individually constructed using the student driving data from Exercise 1.16)

1.19. Given in file NBA2003.MTW are the scoring averages for the top 25 scorers in the National Basketball Association (NBA) in 2002. The data are given below.

	Name		Scoring Average
1	Tracy	McGrady	32.1
2	Kobe	Bryant	30.0
3	Allen	Iverson	27.6
4	Shaquille	O'Neal	27.5
5	Paul	Pierce	25.9
6	Dirk	Nowitzki	25.1
7	Tim	Duncan	23.3
8	Chris	Webber	23.0
9	Kevin	Garnett	23.0
10	Ray	Allen	22.5
11	Allan	Houston	22.5
12	Stephon	Marbury	22.3
13	Antawn	Jamison	22.2
14	Jalen	Rose	22.1
15	Jamal	Mashburn	21.6
16	Jerry	Stackhouse	21.5
17	Shawn	Marion	21.2
18	Steve	Francis	21.0
19	Glenn	Robinson	20.8
20	Jermaine	O'Neal	20.8
21	Ricky	Davis	20.6
22	Karl	Malone	20.6
23	Gary	Payton	20.4
24	Antoine	Walker	20.1
25	Michael	Jordan	20.0

What type of graphical display would you recommend for displaying the data? Construct the display, but before doing so, would you expect the averages to exhibit asymmetry? Why or why not?

Solution:
A dotplot, given below, would be a reasonable way to display the data. We would expect the data to display right skewness as we would expect to see more scoring averages below the median of the top 25 than above the median.

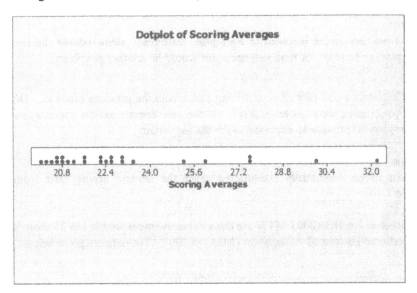

1.21. With a conventional scatter plot, two variables are displayed --- one on the vertical axis and one on the horizontal axis. How many variables were displayed in the scatter plot in Figure 1.5? Can you think of how additional variables might be displayed?

Solution:
Three variables were displayed in Figure 1.5. A fourth variable could be displayed by having a separate graph for each value of that variable. And so on.

1.23. A data set contains 25 observations. The median is equal to 26.8, the range is 62, $Q_1 = 16.7$, and $Q_3 = 39.8$. What is the numerical value of the interquartile range?

Solution:
The interquartile range is $Q_3 - Q_1 = 39.8 - 16.7 = 23.1$

1.25. Would a histogram of the data given in Exercise1.1 be a meaningful display? Why or why not?

Solution:
No, a histogram would not be particularly useful. Since the data are obtained over time, a display that incorporates time should be used.

1.27. What graphical display discussed in this chapter would be best suited for showing the breakdown of the number of Nobel Prize winners by country for a specified time period?

Solution:
A Pareto chart could be a good choice, depending on how many countries are represented. A bar chart would be another possibility, and for political reasons might be the preferred alternative.

1.29. Toss a coin twice and record the number of tails; then do this nine more times. Does the string of numbers appear to be random?

Solution:
(simulation exercise to be performed by student)

1.31. In an article in the Winter, 2001 issue of *Chance* magazine, the author, Derek Briggs found that SAT and ACT preparation courses had a limited impact on students' test results, contrary to what companies that offer these courses have claimed. Read this article and write a report explaining how an experiment would have to be conducted before any claim of usefulness of these courses could be made. (source: http://www.public.iastate.edu/~chance99/141.briggs.pdf)

Solution:
(Student exercise; write a report of the article)

1.33. The following data are frequency distributions of weights of cars and trucks sold in the United States in 1975 and 1990. (*Source:* U.S. Environmental Protection Agency, *Automotive Technology and Full Economic Trends through 1991*, EPA/AA/CTAB/91-02, 1991.)

WT	WT(L)	WT (U)	CA75	TR75	CA90	TR90
1750	1625	1875	0	0	1	0
2000	1875	2125	105	0	109	0
2250	2125	2375	375	0	107	0
2500	2375	2625	406	0	1183	34
2750	2625	2875	281	204	999	45
3000	2875	3250	828	60	3071	428
3500	3250	3750	1029	55	2877	784
4000	3750	4250	1089	1021	1217	1260
4500	4250	4750	1791	386	71	797
5000	4750	5250	1505	201	0	457
5500	5250	5750	828	59	1	46
6000	5750	6250	0	1	0	32

Variable Names:

WT: Weight in pounds, class midpoint
WT(L): Weight in pounds, class lower limit
WT(U): Weight in pounds, class upper limit
CA75: Cars sold, 1975 (thousands)
TR75: Trucks sold, 1975 (thousands)
CA90: Cars sold, 1990 (thousands)
TR90: Trucks sold, 1990 (thousands)

(a) Compare the distributions of CA75 and CA90 by constructing a histogram of each. Comment on the comparison. In particular, does there appear to have been a significant change in the distribution from 1975 to 1990? If so, what is the change? (In MINITAB, the histograms can be constructed using the CHART command with the C1*C2 option; that is, CHART C1 C2 with C1 containing the data and C2 being a category variable, and these two column numbers being arbitrary designations.)

(b) Construct the histograms for TR75 and TR90 and answer the same questions as in part (a).

(c) Having constructed these four histograms, is there any problem posed by the fact that the intervals are not of equal width? In particular, does it create a problem relative to the 1975 and 1990 comparisons? If so, how would you correct for the unequal widths? If necessary, make the appropriate correction. Does this affect the comparison?

(d) In view of the small number of observations, would it be better to use another type of graphical display for the comparison? If so, use that display and repeat the comparisons.

Solution:
(a) The histograms are similar in that they both exhibit extreme right skewness. Beyond that, a finer comparison would not be practical since each histogram is constructed for only 12 observations, so a very large histogram variation would be observed in repeated sampling.

(b) Although both histograms exhibit right skewness, the two large values for TR90 coupled with the fact that there are only 12 observations make it impossible to construct a meaningful histogram.

(c) The unequal widths are a problem with the comparison of TR 75 and TR 90, with the difference in the widths primarily due to the two large values of TR 90.

(d) It would be more meaningful to construct multiple dotplots with the same scale used for each comparison. These effectively show the gaps between the

CA90 and TR90 values, especially in regard to the much smaller differences in the corresponding CA75 and TR75 values, respectively.

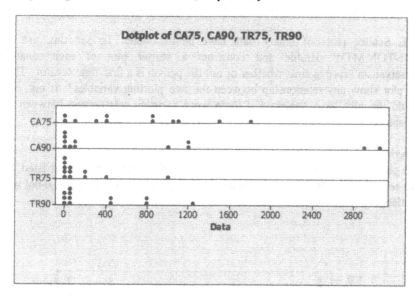

1.35. Use appropriate software, such as MINITAB, or Table A in the back of the book to generate three samples of size 20 from the first 50 positive integers. Compare the three samples. Is there much variability between the three samples? If so, would you have anticipated this amount of variability?

Solution:
(student simulation exercise)

1.37. Consider the Lighthall (1991) article that was discussed at the beginning of the chapter. If you are presently taking engineering courses, can you think of data that should be collected and analyzed on some aspect in an engineering discipline, but that are usually not collected and analyzed? Explain.

Solution:
(student exercise)

1.39. Given the following stem-and-leaf display,

$$3|\ 1\ 2\ 2\ 4\ 5\ 7$$
$$4|\ 1\ 3\ 5\ 7\ 7\ 9$$
$$5|\ 2\ 4\ 5\ 6\ 8\ 9\ 9$$
$$6|\ 1\ 3\ 3\ 4\ 7\ 8$$

determine the median.

Solution:
The median is 52, which is the middle observation among the 25 observations.

1.41. Scatter plots of binary data have limited value. To see this, use the CLINTON.MTW datafile and construct a scatter plot of each senator's conservatism score against whether or not the person is a first-time senator. Does the plot show any relationship between the two plotting variables? If not, how would the plot have appeared if there were a strong relationship between the variables?

Solution:
The plot, given below, shows no relationship. If a relationship had existed, the two sets of vertically aligned points would be close to non-overlapping, if not non-overlapping.

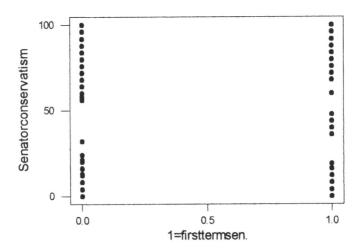

1.43. The file BASKETBALL.MTW contains the NCAA Division 1 highest yearly men's team field goal percentages from 1972 through 2002, which are given below (the years go across the rows):

52.8 52.7 53.0 54.7 53.7 54.5 54.6 55.5 57.2 56.4 56.1
55.6 55.2 54.8 56.1 54.1 54.6 56.6 53.3 53.5 53.6 52.2
50.6 51.7 52.8 52.0 51.8 52.3 50.0 51.1 50.1

(a) Construct either a time series plot of the data or a scatter plot using the year (72, 73, etc.) on the horizontal axis. What type of pattern, if any, would you expect the graph to exhibit? Does the graph appear different from what you expected?

(b) Determine (from an Internet search if necessary) the underlying cause, if any, for the configuration of plotted points. If a cause was discovered, how would you recommend that the graph be reconstructed to show a change?

Solution:

(a) The graph is shown below. There is a clear downward trend starting from 1980. A slightly different configuration might be expected since the three-point field goal was adopted by the NCAA in 1987.

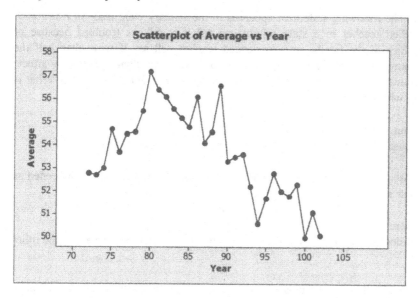

(b) It would be helpful to place a symbol on the graph to indicate the rule change that took effect in 1987.

1.45. What chart would you use (and in fact is used by brokerage companies) to show an investor the breakdown of the total value of his/her stock portfolio into the dollar value of the component parts (plus cash, if applicable)?

Solution:

This is done using a pie chart.

The material in Exercises 46-50 appeared in the "Forsooth" section of *RSS News*, a publication of the Royal Statistical Society, being chosen selectively by the editorial staff from various published sources. The statements therein illustrate mistakes made in the presentation and interpretation of data, and some examples are given in the following problems.

1.47. The following statement appeared in the June 2001 issue of *RSS News*. "The number of official investigations into accidents on building sites is expected to have risen by more than 200 per cent in the last five years while routine inspections have fallen by in excess of 100 per cent over the same period, a Commons written reply showed." (*Original source: The Times*, April 2001.) What is wrong with the statement?

Solution:
It is not possible for any quantity to drop by more than 100%.

1.49. The following statement appeared in the January 2002 issue of *RSS News*: "A skilled teacher in a state school can tell if a child is troubled because of difficulties in the family; an experienced teacher will say if more than half the classroom have a problem, from divorce to drug abuse, then it materially affects the education of the other half." (*Source: Spectator*, June 30, 2001.) What is wrong with this statement?

Solution:
If one part is "more than half", then the other part cannot be "half".

1.51. Why are "percent unfavorable" and "percent favorable" not both needed as columns in a table?

Solution:
Since they add to 100%, if we know one percentage, then we also know the other one.

1.53. The following numbers are grades for a particular class on a test that I gave over 20 years ago: 98, 96, 94, 93, 93, 91, 91, 90, 88, 88, 85, 84, 84, 83, 82, 81, 79, 79, 78, 78, 78, 78, 76, 75, 75, 74, 74, 73, 72, 67, 67, 65, 63, 63, 62, 62, 58, 54, 49, and 44. Indicate what graphical display you would use for each of the following objectives, and then construct each display.

(a) You want to see if there are any large gaps between certain scores. (Of course here the scores are ordered, but numbers won't always be ordered when received.)

(b) You want to obtain an idea of the shape of the distribution of the scores.

Solution:
(a) dotplot (b) histogram

1.55. Assume that someone has constructed a time sequence plot for a particular manufacturing process, but *without* a constant time increment given on the horizontal axis. If a time sequence plot for a similar manufacturing process (same product) was also constructed, could the two plots be meaningfully compared? Explain.

Solution:
The plots could be compared since they were constructed the same way, but either plot alone could be misleading because of the non-constant time increment.

1.57. Do finite populations occur and are they of interest in your field of engineering or science? If so, give three examples; if not, explain why they do not occur.

Solution:
(student exercise)

1.59. Consider the 75th percentile SAT scores for the data given in Exercise 1.7. What would you expect for the shape of a histogram of these scores? Construct a histogram of the scores using 8 classes. Is the shape what you expected?

Solution:
We might expect to see some skewness resulting from the very high 75th percentile scores of the schools at the top of the rankings. The histogram with 8 intervals is given below.

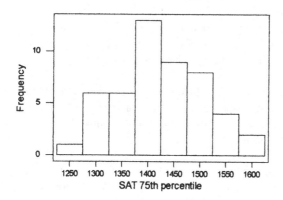

The shape is not surprising.

1.61. To illustrate a stem-and-leaf display that is more sophisticated than the one given in Section 1.4.1, use MINITAB (or other software) to generate 100 random integers between 1 and 10, with the integers all being equally likely to be selected. The MINITAB command for a stem-and-leaf display is STEM. Use that (or other software) to produce a stem-and-leaf display of these 100 random integers. Is the appearance of the display unexpected? Explain. If MINITAB was used, interpret the numbers to the left of the display.

Solution:
Given below is one such stem-and leaf display

```
Stem-and-leaf of C8     N = 100
Leaf Unit = 0.10

 12    1 000000000000
 20    2 00000000
 32    3 000000000000
 42    4 0000000000
(10)   5 0000000000
 48    6 00000000
 40    7 000000000
 31    8 0000000000000
 18    9 0000000000000
  5   10 00000
```

The numbers to the left of the integers are "depth" counts (i.e., cumulative frequencies), starting from each end, with the median indicated as the number with the frequency in parentheses.

1.63. The distribution of grade-point averages (GPAs) for all 1,999 sorority members during the second semester of the 2001-2002 academic year at Purdue University was shown by a histogram to be highly left-skewed.

(a) Five classes were used for the histogram. Would you suggest that more or fewer classes be used in an effort to present a clear picture of the shape of the distribution? If you believe that a different number should be used, how many class intervals would you suggest?

(b) With the highest GPA being 3.222, the lowest 2.538, and Q_1 and Q_3 being 3.148 and 2.839, respectively, would you expect the distribution to still be left-skewed if a different number of classes had been used? Explain.

(c) It was stated in Section 1.5.1 that data are generally not left-skewed, and indeed the distribution of GPAs for members of a fraternity that semester was close to being symmetric. In particular, the highest GPA was 3.130 and the value of Q_3 was 2.7535, with the difference between these two numbers being much greater than for sorority members. What graphical device would you use (which was not used at the Purdue website) to show the difference in the distributions for fraternity and sorority members, especially the difference in skewness?

Solution:

(a) Far more than five classes should be used. A reasonable number would be 11, which is the number suggested by the power-of-two rule.

(b) Left-skewed data are rare, so that should have been a tipoff that the histogram with five classes was misleading. We would logically expect the data to be right skewed, and the information given here, albeit limited, also suggests that.

(c) One possibility would be to construct curves for each distribution from the separate histograms and then overlay the curves.

1.65. Consider the statistics for teams in the National Basketball Association (NBA) that are available at http://www.nba.com/index.html and at

http://sports.espn.go.com/nba/statistics?stat=teamstatoff&season=2003&seasonty pe=2

(a) Construct a scatter plot of overall field goal percentage versus 3-point field goal percentage for the 29 teams. Would we expect the scatter plot to show a relationship? Why or why not?

(b) Construct a scatter plot of 3-point field goal percentage versus 2-point percentage, after performing the appropriate calculations to obtain the 2-point percentages. Does the scatter plot show a relationship, and if so, is the relationship expected?

(c) Construct a boxplot of the average number of 3-point field goal attempts per game for the 29 teams. Does the boxplot show any outlier teams? Explain.

Solution:

(a) The scatterplot, shown below, does indicate a moderate correlation between the two percentages, and certainly this should occur since the overall percentage is a weighted average of the 3pt. percentage and the 2pt. percentage.

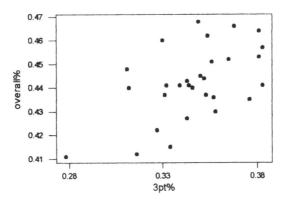

(b) Interestingly, there is essentially no relationship, as the appearance of a weak relationship is due to the outlying observation in the lower left part of the graph.

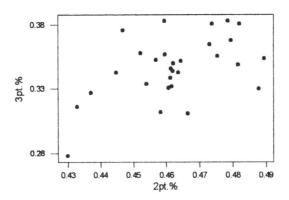

(c) The box plot, given below, shows one outlier: Boston at 26.3, which is much greater than the second-highest value, Dallas at 20.3.

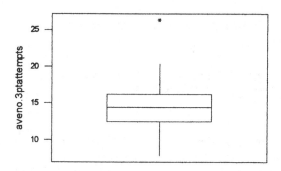

1.67. What can a time sequence plot of individual observations reveal that could be at least partially hidden in a time sequence plot of averages of 5 observations (as in quality control work)?

Solution:
A time sequence plot would show whether or not there are any extreme observations that could be hidden by the averages. If the variability of a process is very much out of control, extreme observations in opposite directions could theoretically occur and yet the average could be about what would be expected.

1.69. Name three engineering continuous random variables that you would expect to result in a moderately skewed distribution if 100 values were obtained on each variable. Name three discrete random variables that relate to a company's clerical operations that you would also expect to be skewed. Should any of these variables be left-skewed (tail to the left) or should they all be right-skewed?

Solution:
(student input exercise)

1.71. The Browser Summary report for hits to the U.S. Geological Survey website for November 2002 showed Internet Explorer accounted for 69.59% of the hits, 7.30% for Netscape, 4.14% for Inktomi Search, and 4.13% for Googlebot, in addition to the percentages for 29 other browsers. What graphical display given in this chapter would seem to be best suited for displaying this information?

Solution:
Pareto chart

1.73. Consider the data given in Exercise 1.7. Plot the 75th percentile scores against the 25th percentile scores and then plot the 25th percentile scores against the 75th percentile scores.

(a) Compare the configuration of points and comment. Would you expect the plots to differ by very much? Explain.

(b) In general, under what conditions would you expect a plot of Y versus X to differ noticeably from a plot of X versus Y?

Solution:
(a) Of course the plots are mirror images. We would expect the plots to be quite similar since we would expect a high correlation between the plotted points so that something very close to a straight-line configuration should result for each plot. Therefore, even though the plots are mirror images, they should not differ greatly.

(b) The extent to which the plots differ will be a function of the extent to which either of them differs from a straight line.

1.75. Would you recommend that a pie chart be constructed with 50 "slices"? Why or why not? In particular, assume that there are 50 different causes of defects of a particular product. If a pie chart would not be suitable for showing this breakdown, what type of graph would you recommend instead?

Solution:
A pie chart with 50 slices would be virtually impossible to read. A Pareto chart should be constructed instead.

1.77. The Boise Project furnishes a full irrigation water supply to approximately 400,000 acres of irrigable land in southwestern Idaho and eastern Oregon (see http://www.usbr.gov/dataweb/html/boise.html). The following data are from that website.

Year	Actual Area Irrigated (Acres)	Crop Value (Dollars)
1983	324,950	168,647,200
1984	327,039	149,081,226
1985	325,846	135,313,538
1986	320,843	151,833,166
1987	309,723	153,335,659
1988	308,016	157,513,694
1989	326,057	185,990,361
1990	323,241	177,638,311
1991	323,241	177,145,462
1992	325,514	182,739,499

What graphical display (if any) would you use to determine if there seems to be a relationship between area irrigated and crop value over time? Does there appear to be a relationship between the two?

Solution:
The scatterplot, given below, does not show any relationship between the two.

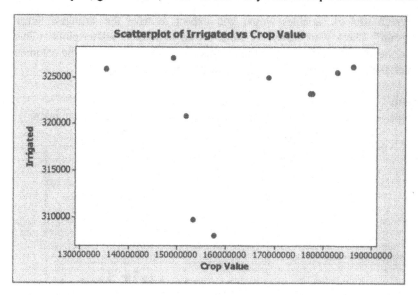

1.79. One of the sample datafiles that comes with MINITAB, MASSCOLL.MTW in the STUDNT12 subdirectory of datafiles, contains various statistics for the 56 four-year colleges and universities in Massachusetts in 1995. Consider the variable Acceptance Rate, with the following data:

```
%Accept
 76.9718    23.0590    80.8571    79.6253    56.7514
 67.2245    76.7595    47.1248    64.4166    74.4921
 65.5280    67.9552    71.3197    56.4796    76.4706
 79.2636    84.8980    66.4143    85.8000    73.1537
 55.9285    83.8279    74.3666    15.6149    87.3786
 81.1475    33.3801    77.7890    73.0000    85.0737
 90.0000    64.2994    71.3553    91.2429    90.1639
 79.7382    77.9661    57.0000    54.6325    71.3453
 63.0828    78.5886    47.3470    85.9814    66.8306
 77.5919    75.9157    43.1433    84.9324    89.1515
 69.3548    78.9989    83.9599    29.7420    83.5983
 68.9577
```

If you wanted to show the distribution of this variable over the 56 schools and spotlight extreme observations, what graphical technique would you employ? Produce the graph and comment. Over the years there have been suspicions (not necessarily for Massachusetts) that self-reporting of certain statistics such as average SAT score has resulted in exaggerated numbers being reported.

(a) Recognizing that there are many different variables in this datafile, how would you check to determine if any of the numbers looked out of line?

(b) Which pair of variables would you expect to have the strongest linear relationship? Check your answer by constructing all possible scatter plots. (This can be done in MINITAB by using the command MATR followed by the columns for which the set of scatter plots is to be constructed.

Solution:
A dotplot would be a reasonable type of graph to use. The plot is shown below.

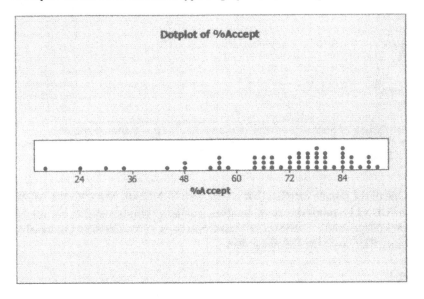

(a) Any numbers that are out of line are probably more likely to be out of line in a multi-dimensional way than in a one-dimensional way. Therefore, scatterplots could be revealing. Given below is the scatterplot of MSAT score against Acceptance Rate, which does show one suspicious-looking point: the point with a relatively high MSAT average that is well-removed from the linear relationship exhibited by the other points.

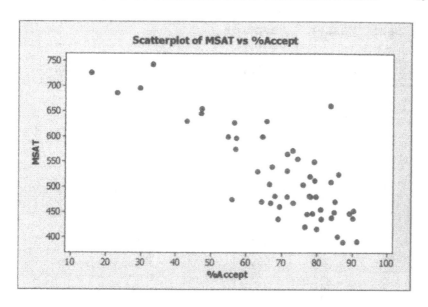

(b) We would expect the strongest linear relationships to occur among the aptitude test scores, and that is shown in the matrix scatterplot below.

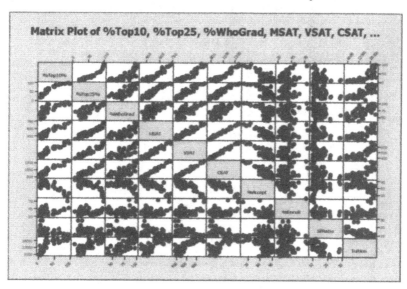

1.81. One of the sample datafiles that comes with MINITAB is GAGELIN.MTW. The dataset is from the *Measurement Systems Analysis Reference Manual* (Chrysler, Ford, General Motors Supplier Quality Requirements Task Force). Five parts were selected by a plant foreman that covered the expected range of measurements and each part was randomly measured 12 times by an operator. The data are given below.

Row	Part	Master	Response
1	1	2	2.7
2	1	2	2.5
3	1	2	2.4
4	1	2	2.5
5	1	2	2.7
6	1	2	2.3
7	1	2	2.5
8	1	2	2.5
9	1	2	2.4
10	1	2	2.4
11	1	2	2.6
12	1	2	2.4
13	2	4	5.1
14	2	4	3.9
15	2	4	4.2
16	2	4	5.0
17	2	4	3.8
18	2	4	3.9
19	2	4	3.9
20	2	4	3.9
21	2	4	3.9
22	2	4	4.0
23	2	4	4.1
24	2	4	3.8
25	3	6	5.8
26	3	6	5.7
27	3	6	5.9
28	3	6	5.9
29	3	6	6.0
30	3	6	6.1
31	3	6	6.0
32	3	6	6.1
33	3	6	6.4
34	3	6	6.3
35	3	6	6.0
36	3	6	6.1
37	4	8	7.6
38	4	8	7.7
39	4	8	7.8
40	4	8	7.7
41	4	8	7.8
42	4	8	7.8
43	4	8	7.8
44	4	8	7.7
45	4	8	7.8
46	4	8	7.5
47	4	8	7.6
48	4	8	7.7
49	5	10	9.1
50	5	10	9.3
51	5	10	9.5
52	5	10	9.3
53	5	10	9.4
54	5	10	9.5
55	5	10	9.5
56	5	10	9.5

57	5	10	9.6
58	5	10	9.2
59	5	10	9.3
60	5	10	9.4

What graphical device would you use to show the variability in measurements by the operator relative to the reference value for each part that is given in the adjacent column? Be specific. Would a histogram of the measurements show anything of value? Explain.

Solution:

A scatterplot of the difference between the measurement and the master value against the master value shows the variability relative to the latter. The plot of this given below exhibits two extreme differences for the master value of 4, but otherwise the variability in the differences does not differ greatly over the master values. A histogram of the measurements would not provide any useful information because the master values are not constant.

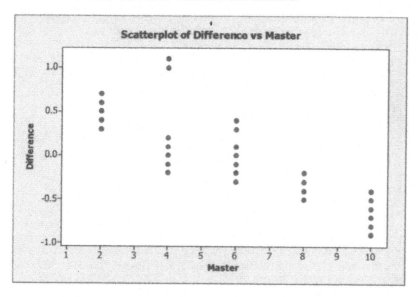

1.83. It was stated in Exercise 1.1 that there were two errors in the source from which the data was obtained. For that type of data, what method(s) would you use to try to identify bad data points?

Solution:
A time series plot can be used to detect bad data collected over time, and in fact that is how the second bad data point was detected.

2

Measures of Location and Dispersion

2.1. The following are six measurements on a check standard, with the check standard measurements being resistivities at the center of a 100 ohm.cm wafer: 96.920, 97.118, 97.034, 97.047, 97.127, and 96.995.

(a) Compute the average and the standard deviation of the measurements.

(b) Could the average have easily been computed without having to enter the numbers as given in a computer or calculator (e.g., by entering integers in a calculator)? Explain.

(c) Similarly, assume that one wished to compute the variance without using the numbers as given. Could this be done? Explain.

Solution:

(a) The average is 97.040 and the standard deviation is 0.078.

(b) The average could be easily computed without a calculator since the numbers all deviate only slightly from 97. Treat the decimal fraction part as integer deviations from 97 (− 80, 118, 34, etc.). Then add the deviations sequentially in your head, divide by 6, then divide by 1000 and add the result to 97. This produces 97 + (241/6)/1000 = 97.0402. With practice, calculations of this sort can be performed faster than by punching numbers in a calculator or computer. (I know because I did so for many years.)

(c) Using a similar approach to compute the variance would not be practical for this example, due in part to the fact that the average is not an integer.

2.3. A "batting average" was mentioned in Section 2.1 as being one type of average. What is being averaged when a batting average is computed? Is this what sportswriters and baseball fans think about being averaged to obtain a batting average? Explain.

Solution:

A batting average is actually an average of ones and zeros, with a one being assigned when a hit is recorded and a zero assigned when an out is made. It is unlikely that very many baseball fans know what is actually being averaged to obtain a batting average.

2.5. Show by appropriate substitution in the expression for σ that the value of σ for heights in inches for some population is 12 times the value of σ when height is measured in feet (and fractions thereof).

Solution:
Height in inches is 12 times height in feet (e.g., 5-9 is 5.75 feet and 12(5.75) = 69 inches) The factor of 12 becomes squared when substituted in the expression for the variance, but then becomes 12 when the standard deviation is computed by taking the square root of the variance.

2.7. We can always compute an average of a group of numbers but we also need to ask whether or not the average will make any sense. For example, the following numbers are provided by Western Mine Engineering, Inc. (www. westernmine.com); they represent the operating cost indices for surface mines in the United States from 1988 to 2000: 85.5, 90.4, 95.7, 97.7, 98.7, 99.0, 100.0, 102.4 105.7, 107.0, 106.7, 108.7, 113.5.

(a) Compute the average of these 13 numbers.

(b) Does this number estimate some population parameter? If so, what is it?

(c) Compute the standard deviation of the numbers. What is the unit of measurement for the standard deviation?

Solution:
(a) The average is 100.85.

(b) There is no parameter that is being estimated with this average.

(c) The standard deviation is 7.69. The unit of measurement for the standard deviation is the original unit of measurement: cost index.

2.9. Show that $\sum_{i=1}^{n}(X_i - \overline{X})^2 = \sum_{i=1}^{n} X_i^2 - (\sum_{i=1}^{n} X_i^2)^2/n.$

Solution:
$$\sum_{i=1}^{n}(X_i - \overline{X})^2 = \sum_{i=1}^{n}(X_i^2 - 2\overline{X}X_i + \overline{X}^2) = \sum_{i=1}^{n} X_i^2 - 2\overline{X}\sum_{i=1}^{n} X_i + \sum_{i=1}^{n}\overline{X}^2 =$$
$$\sum_{i=1}^{n} X^2 - 2n\overline{X}^2 + n\overline{X}^2 = \sum_{i=1}^{n} X^2 - n\overline{X}^2 = \sum_{i=1}^{n} X_i^2 - (\sum_{i=1}^{n} X_i^2)^2/n.$$

2.11. For the data described in Exercise 2.10, assume that you want to construct a boxplot to summarize the salaries of starting pitchers. You are not given the raw data but you are given 10 summary statistics.

(a) Which statistics do you need in order to construct the boxplot?

(b) Would a boxplot be an appropriate choice for a graphical display given the number of observations and the nature of the data?

Solution:

(a) The median, 75th percentile, 25th percentile, and largest and smallest values must be known in order to construct a skeletal boxplot.

(b) Yes, a box plot would be an appropriate choice.

2.13. Let k denote the standard deviation for a sample of 25 numbers.

(a) If each number is multiplied by 100, what is the standard deviation of the new set of numbers, as a function of k?

(b) Is the coefficient of variation affected by this multiplication? Explain.

Solution:

(a) The standard deviation for the new numbers is 100 times the standard deviation of the original numbers.

(b) The coefficient of variation is unaffected because the mean of the new numbers is also 100 times the mean of the original numbers.

2.15. Assume that you work for a company with 500 employees and you have to fill out a report on which you will give the typical salary for an employee with the company. What statistic will you use for this purpose? Would your answer be any different if the company had 50 employees, or 5,000 employees?

Solution:

A typical salary is best represented as the median salary. The need to use the median instead of the average shouldn't depend greatly on company size since the larger the company, the more personnel there are with high salaries.

2.17. Many instructors compute summary statistics on test scores and provide these to the class or post the results outside the instructor's office door. If the variance of the test scores is given, what is the unit for that measure? Would you recommend that an instructor use the variance or standard deviation as the measure of variability, or perhaps some other measure?

Solution:

The unit for the variance would be the test score squared. It would be better to use the standard deviation so that the students could see where they stood in terms of the number of standard deviations that their scores are from the average.

2.19. Assume that a random variable is measured in inches. Which of the following will *not* be in inches: (a) mean, (b) variance, and (c) standard deviation?

Solution:

The variance will not be in inches since the unit is always the square of the original unit.

2.21. A sample of 50 observations is modified by dividing each number by 3 and then subtracting 75. That is, $Y = X/3 - 75$, with X representing the original observations. If $s_Y^2 = 1.2$, what is s_X^2?

Solution:
Solving for X, we obtain $X = 3Y + 225$. Therefore, $s_X^2 = (3)^2 s_Y^2 = 9(1.2) = 10.8$.

2.23. Purchasing shares of company stocks became affordable (in terms of commissions) for the average person some years ago with the advent and popularity of online brokerages. If an investor buys 200 shares of JBL at 28.40 in December, 100 shares at 27.64 in January, and 150 shares at 25.34 in March, (a) what was the total amount that the investor paid for the stock, and (b) what was the average price paid for the 450 shares that were purchased? If 200 additional shares were bought in June at that average price, would the average for the 650 shares change? Why or why not?

Solution:
(a) 200(28.40) + 100(27.64) + 150(25.34) = $12,245 = total amount paid.

(b) Cost per share is 12,245/450 = $27.21. The cost would remain unchanged if an additional 200 shares were purchased at this price since the two components in the weighted average would each be $27.21.

2.25. Once when I was a graduate student, a fellow graduate student wanted to know the average grade he assigned for a particular semester. Assume that the grades were 14 A's, 24 B's, 26 C's, 10 D's, and 4 F's. (This was over 30 years ago, in the days before grade inflation.) Computing the overall average in one's head, as I did, is a type of "lightning calculation", as it is called. Letting A, B, C, D, and F be designated by 4, 3, 2, 1, and 0, respectively, compute the average grade assigned in your head if you can do so; otherwise, use a calculator.

Solution:
Doing the computations in one's head should show that the average is 2.44.

2.27. For a normal distribution (see Section 3.4.3), the variance of S^2 is $2\sigma^4/(n-1)$. Thus, if σ^2 is much greater than, say, 20, the variance of S^2 will be greater, for practical sample sizes, than the parameter that the statistic is trying to estimate. What does this suggest about the choice of sample size when the population variance is believed to be large?

Solution:
A very large sample should be used unless a large estimation error can be tolerated.

2.29. Assume two data sets, with the second one obtained by multiplying each number in the first data set by 0.9. If the coefficient of variation of the first data set is a, what is the coefficient of variation for the second data set as a multiple of a?

Solution:
The coefficient of variation of the second data set is also "a", so the multiple is 1. The fact that the numbers are multiplied by a number less than 1 does not have a distorting effect on the coefficient of variation as the number is unaltered regardless of the value of the multiplier, as long as the multiplier is positive.

2.31. The following definition of a statistic is given at an Internet site: "any number calculated from sample data, describes a sample characteristic." Can we accept this definition and still speak of the variance of a statistic? Explain.

Solution:
Technically, a statistic is some function of the data in a sample. It is not a number; which is the numerical value of a statistic.

2.33. An employee needs to perform a quick calculation to find the average of 50 numbers, all of which are of the form 6.32xx. To simplify the computation, she visually multiplies each number by 10,000 and then subtracts 63,200 from the transformed number, thus leaving the "xx" part in integer form. If the average of the transformed numbers is 42.8, what is the average of the original numbers?

Solution:
Let X = original number. Then $Y = 10{,}000X - 63{,}200$
$\Rightarrow X = (Y + 63{,}200)/10{,}000$ so $\overline{X} = (\overline{Y} + 63{,}200)/10{,}000 = (42.8 + 63{,}200)/10{,}000 = 63{,}242.8/10{,}000 = 6.32428$.

2.35. Assume that $n = 40$ and $\sum_{i=1}^{n}(x - a)^2 = 612$. If $a = \overline{x} + 3$, what is the numerical value of $\sum_{i=1}^{n}(x - \overline{x})^2$?

Solution:
$\sum(x - a)^2 = 612 \rightarrow \sum(x - (\overline{x} + 3)^2) \rightarrow \sum((x - \overline{x}) - 3)^2 \rightarrow$
$\sum(x - \overline{x})^2 + \sum 9$ (the middle term vanishes when the binomial is squared since
$\sum((x - \overline{x}) = 0)$. Thus, $\sum(x - \overline{x})^2 + 40(9) = 612 \Rightarrow \sum(x - \overline{x})^2 = 252$.

2.37. The sample variance (s^2) is found to be 35.6 for a sample of 25 observations. If each of the original numbers were multiplied by 20 and then 10 was subtracted from each of the transformed numbers, what would be the value of s^2 for the transformed set of 25 numbers?

Solution:
The variance of the transformed numbers would be $20^2(35.6) = 14,240$.

2.39. Which would be more adversely affected by outliers, the sample standard deviation, s, or a statistic defined as $\sum_{i=1}^{n}|x_i - \bar{x}|/\sqrt{n}$? Explain. Which statistic would you recommend for general use?

Solution:
A statistic like $\sum_{i=1}^{n}|x_i - \bar{x}|/\sqrt{n}$ would be less affected by outliers since a large deviation would not be squared. Statistics such as this would have a disadvantage in that they are not part of commonly used statistical procedures, however.

2.41. Assume that we have 100 numbers and subsequently put them in ascending order.

(a) Explain how the median would be computed.

(b) If each of these 100 ordered numbers were multiplied by 1.5, what would be the relationship between the median of these transformed numbers and the median of the original numbers?

Solution:
(a) The median would be the average of the 50th and 51st ordered observations.

(b) The median will be 1.5 times the median of the original numbers.

2.43. Construct a sample of size 3 for which the sample variance exceeds the mean.

Solution:
One such example would be 4, 14, and 24, as the variance is 100 and the average is 50. (Note that the standard deviation, 10, is equal to the average deviation between the ordered numbers. It is possible to construct many such examples for which this is the case.)

2.45. A set of data is transformed to make the data easier to work with. If the sample variance for the original data was 125.67 and the sample variance of the transformed data was 1.2567, what transformation was used? Could there be more than one answer to this question? Explain.

Solution:
The transformation must have been of the form $Y = 0.1X + c$, with c an arbitrary constant. Thus, there is essentially an infinite number of possible transformations as c could be any real number.

2.47. Assume you are given a sample of six numbers that are deviations of each original number from the average of the numbers. One of the numbers was given with the wrong sign, however. If the deviations are 2.3, 3.1, 4.2, $- 5.2$, and 1.8, which number has the wrong sign?

Solution:
The number must be half of the sum of the deviations. The sum is 6.2 so the deviation that has the wrong sign is 3.1, as it should be $- 3.1$.

2.49. Determine the median from the following stem-and-leaf display (given in Section 1.4.1):

   6| 4 5

   7| 2 2 3 4 8 9

   8| 2 3 5 7 7 8

   9| 2 3 6 7

Solution:
The median is 84 (the average of 83 and 85).

2.51. If the standard deviation of a set of 50 numbers is 6.23, what will be the standard deviation of the transformed numbers if each number is multiplied by 100 and then 600 is subtracted from each number that results from the multiplication?

Solution:
The standard deviation of the new numbers is 100 (6.23) = 623.

2.53. Assume that an engineering statistics class consists of 25 men and 10 women. If the median height of the women is 65.4 inches and the median height of the men is 71.6 inches, can the median height for the men and women combined be determined? Why or why not?

Solution:
No, not enough information is given.

2.55. Consider the following sample of 24 observations: 10, 12, 13, 15, 18, 21, 24, 27, 29, 32, 33, 35, 36, 38, 42, 44, 45, 46, 48, 49, 51, 52, 53, and 56. Compute the mean and the standard deviation of the ungrouped data. Then put the data in classes of 10-19, 20-29, 30-39, 40-49, and 50-59 and compute the estimated mean and estimated standard deviation from the grouped data. Compare the results and comment.

Solution:
The mean and standard deviation for the ungrouped data are 34.54 and 14.43, respectively. The mean for the grouped data is 34.50 and the standard deviation without the correction for grouping is 14.14. The standard deviation with the correction for grouping would be less than 14.14, so in this case the correction would be in the wrong direction.

2.57. Cornell [*ASQC Statistics Division Newsletter*, **14**(1), 11-12, 1994] explained, in a lighthearted article, how a person could maximize the octane rating of the gas in his or her tank for a given cost, or conversely, minimize the cost for a given octane rating. Sample the prices for three service stations in your neighborhood and for each service station determine the mixture of 87, 89, and 92 (or 93) octane that should be put in a car to maximize the octane rating for a $15 purchase. (Assume for the sake of the illustration that a mixture of different octanes wouldn't harm the engine.) Then determine how to obtain an octane rating of 88 for the lowest cost. Note that the problem involves weighted averages.

Solution:
(student exercise)

2.59. The American Statistical Association conducts salary surveys of its members. The results are published and are also available at the ASA website http://www.amstat.org/profession/salarysurvey.pdf. Assume that you want to analyze the variability in the salary of full professors at research universities and you want to see how the variability changes over time (i.e., over surveys). The individual salaries are not given, however, so a variance or standard deviation could not be computed. Furthermore, the class intervals (i.e., for the number of years at the rank of full professor) are unequal and the last class is open-ended (e.g., 33 or more years). The frequencies of each class are given, in addition to Q_1, Q_2, and Q_3. Given this limited information, how would you compare the variability in the salaries over time?

Solution:
The sample variance for grouped data could be computed, but without using the correction factor since the class widths are unequal. Although the last class is open-ended, it would be reasonable to use 35 as the midpoint. Furthermore, because of the relatively large number of classes, the choice of a value for the midpoint is not apt to make much difference. Separate variances for the median, Q_1, and Q_3 might be computed and the results compared.

2.61. As an exercise, a student computes the average of three numbers. One number is two units above the average and another number is equal to the average. What is the standard deviation of the three numbers?

Solution:
The standard deviation is 2. (The third number must be two units below the average.) This is one of those rare examples for which the standard deviation is equal to the average deviation between consecutive ordered numbers. Although the two statistics will hardly ever be equal in practice, they should be of the same order of magnitude.

2.63. Obtain the most recent salary information for professionals in your major field and, assuming it is in the form of class intervals with corresponding class frequencies, describe the distribution. Is the distribution at least close to being symmetric in regard to years in the profession? Explain.

Solution:
(data-gathering exercise performed by student)

2.65. Consider the (ordered) sample observations: 2, 8, 23, 24, 28, 32, 34, 38, 42, 44, 45, 55, 58, 61, 65, 68, 71, 72, 98, and 99. Compute the sample variance with 10% trimming (from each end), and compare this number with the sample variance computed without trimming. Comment.

Solution:
The variance of all the numbers is 691.6; the variance after 10% trimming is 288.1. The numbers differ greatly because of the extreme observations.

2.67. The average of 10 numbers is 100. If the numbers 79 and 89 are removed from the group of numbers and the average of the eight remaining numbers is computed, what is the average of those numbers?

Solution:
The sum of the numbers without the 79 and 89 is $10(100) - 79 - 89 = 832$. Thus, the average is $832/8 = 104$.

2.69. The following statistics were given at the website for the U.S. Geological Survey (USGS), with the numbers pertaining to the USGS External Server and the statistics being for one particular month during the last quarter of 2002.

Successful requests	14,744,979
Average successful requests per day	491,522
Successful requests for pages	1,471,649

Average successful requests for pages per day   49,057

(a) Can the month for which the statistics were given be determined? If so, what is the month? If the month cannot be determined, explain why not.

(b) The averages are obviously rounded to the nearest integer. What was the units digit in each number before they were each rounded, if this can be determined? If it cannot be determined, explain why.

Solution:
(a) It must have been November because the appropriate division shows that the month had 30 days, although the numbers seem to be slightly off.

(b) Actually this cannot be determined because the numbers are obviously not correct within rounding.

2.71. Can a sample of size 2 be constructed for which the standard deviation and the variance are the same? If so, construct the sample. If it is not possible, explain why it can't be done.

Solution:
This can occur in only two ways: in the trivial case where both the numbers are the same so that the standard deviation and variance are both zero, and in any of an infinite number of examples for which the variance is one. One such example is $3 + \sqrt{0.5}$, $3 - \sqrt{0.5}$.

2.73. Construct a sample of 10 observations for which the range is twice the interquartile range.

Solution:
One sample of 10 observations would be 10, 12, 14, 16, 18, 20, 22, 24, 26, and 32, which has an interquartile range of 11 (with the method that MINITAB uses for computing quartiles), and a range of 22.

2.75. The average salary paid to all employees in a company is $50,000. The average annual salaries paid to male and female employees were $52,000 and $42,000, respectively. Determine the percentage of males and females employed by the company.

Solution:
Let a = percentage of men. So, $a(52,000) + (1 - a)(42,000) = 50,000$. Solving the equation produces $a = .80$. Thus, 80% of the employees are men and 20% are women.

3

Probability and Common Probability Distributions

3.1. What is the numerical value of P(B|A) when A and B are mutually exclusive events?

Solution:
P(B|A) = 0 because B cannot occur if A has occurred since the events are stated as being mutually exclusive.

3.3. A survey shows that 40% of the residents of a particular town read the morning paper, 65% read the afternoon paper, and 10% read both.

(a) What is the probability that a person selected at random reads the morning or the afternoon paper, or both?

(b) Now assume that the morning and afternoon papers are subsequently combined, as occurred with the Atlanta papers in February 2002, with the paper delivered in the morning. If the paper doesn't lose (or gain) any subscribers, can the probability that a person selected at random receives the single paper be determined? If so, what is the probability? If not, explain why not.

Solution:
(a) The probability that at least one of the two papers is read is $.60 + .45 - .10 = .95$, using the rule that $P(A \cup B) = P(A) + P(B) - P(A \cap B)$.

(b) This would be the probability that a person reads at least one of the papers so that the person would thus be reading the combined paper, which as in part (a), is also .95.

3.5. Given $P(A) = 0.3$, $P(B) = 0.5$, and $P(A \mid B) = 0.4$, find $P(B \mid A)$.

Solution:
$P(B)P(A|B) = P(A \cap B) = .5(.4) = .2$. $P(B|A) = P(B \cap A)/P(A) = .2/.3 = 2/3$

3.7. Give an example of a discrete random variable and indicate its possible values. Then assign probabilities so as to create a probability mass function.

Solution:
A weather forecaster gives his probabilities for conditions at 6 p.m. the next day for the city in which he works, with light rain assigned a probability of .3, heavy rain .2, overcast .4, and clear .1. That is his probability distribution.

3.9. Given the probability density function $f(x) = 4x$, $1 < x < \sqrt{1.5}$, determine $P(1.0 < X < 1.1)$.

Solution:
$$P(1.0 \leq X \leq 1.1) = \int_1^{1.1} 4x \, dx = .42.$$

3.11. Assume that the amount of time a customer spends waiting in a particular supermarket checkout lane is a random variable with a mean of 8.2 minutes and a standard deviation of 1.5 minutes. Suppose that a random sample of 100 customers is observed.

(a) Determine the approximate probability that the average waiting time for the 100 customers in the sample is at least 8.4 minutes.

(b) Explain why only an approximate answer can be given.

Solution:

(a) $P(\overline{X} \geq 8.4)$: $Z = \dfrac{\overline{X} - \mu}{\frac{\sigma}{\sqrt{n}}} = \dfrac{8.4 - 8.2}{\frac{1.5}{\sqrt{100}}} = 1.33$ $P(Z \geq 1.33) = .0918.$

(b) The answer in part (a) is just an approximation since normality was not assumed and the Central Limit Theorem must thus be invoked.

3.13. A balanced coin is tossed six times. What is the probability that at least one head is observed?

Solution:
$P(\text{at least one head}) = 1 - P(\text{zero heads}) = 1 - (1/2)^6 = 1 - 1/64 = 63/64.$

3.15. Explain how you might proceed to determine a continuous distribution that provides a good fit to your set of data.

Solution:
There are various ways to proceed, including using software for automated fitting. Alternatively, the literature can be suggestive of one or more distributions for certain types of random variables (roundness, flatness, etc.). Finally, the first four sample moments might be computed so as to try to match these values with distributions whose moments are close to the sample moments.

3.17. Explain what is wrong with the following statement: "The possible values of a binomial random variable are 0 and 1."

Solution:
The possible values of a binomial random variable are 0 through n. The possible values for each Bernoulli trial are 0 and 1.

3.19. Quality improvement results when variation is reduced. Assume a normal distribution with a known standard deviation, with 34% of the observations contained in the interval $\mu \pm a$. If the standard deviation is reduced by 20% but the mean is not kept on target and increases by 10% from μ, obtain if possible the percentage of observations that will lie in the interval $\mu \pm a$? If it is not possible, explain why the number cannot be obtained.

Solution:
It is not possible to solve for this percentage since μ is unknown.

3.21. Show that $E(a + bX) = a + bE(X)$ and $Var(a + bX) = b^2 Var(X)$.

Solution:
$E(a + bX) = E(a) + E(bX) = a + bE(X)$
$Var(a + bX) = Var(bX) = b^2 Var(X)$ since an additive constant does not affect the variance but a multiplicative constant does so.

3.23. A company employs three different levels of computer technicians, which it designates with classifications of A, B, and C. Thirty percent are in classifications A and B, and 40% are in C. Only 20% of the "A" technicians are women, whereas the percentages for B and C are 30% and 35%, respectively. Assume that one technician has been selected randomly to serve as an instructor for a group of trainees. If the person selected is a man, what is the probability that he has a B classification?

Solution:
This is a Bayes Theorem problem. Let A, B, and C denote the classifications and let M and W, denote men and women, respectively. We want $P(B \mid M)$.

$$P(B|M) = \frac{P(B)P(M|B)}{P(M)} \text{ with } P(M) = P(A)P(M|A) + P(B)P(M|B) + P(C)P(M|C).$$
$$= \frac{(.3)(.7)}{(.3)(.8) + (.3)(.7) + (.4)(.65)} = \frac{.21}{.71} = .296$$

3.25. Assume that you use a random number generator to generate an integer from the set of integers 1-4 and the selection is truly random. Let X represent the value of the selected integer and compute $Var(X)$.

Solution:
$f(x) = 1/4$, $x = 1, 2, 3, 4$. $E(x) = 2.5$ (by inspection since it must be the simple average of the four possible values). Similarly, $E(x^2) = 7.5$, the average of the possible x^2 values. Then $Var(x) = 7.5 - (2.5)^2 = 7.5 - 6.25 = 1.25$.

3.27. If a die is rolled three times, give the probability distribution for the number of twos that is observed.

Solution:

The appropriate distribution is the binomial distribution (i.e., twos and not twos), which is used to produce the following probability distribution for the number of twos in three rolls of the die:

X	$P(X)$
0	125/216
1	75/216
2	15/216
3	1/216

3.29. You are contemplating investing in a particular stock tomorrow and one of your brokers believes there is a 40% chance that the stock will increase over the next few weeks, a 20% chance that the price will be essentially unchanged, and a 40% chance that the price will decline. You ask another broker and she gives you 35%, 30%, and 35%, respectively, for the three possible outcomes. Assume that you accept these percentages. If X = the outcome under the first broker's judgment and Y = the outcome under the second broker's judgment, which random variable has the smaller variance? What course of action would you follow?

Solution:
Let the possible outcomes be denoted by 0, 1, and 2. The choice is arbitrary; the variance does not depend upon the choice. Using the corresponding probabilities for each possible outcome for the first broker and the second broker, it can be shown that $Var(X) = 0.8$ and $Var(Y) = 0.7$. Since $E(X) = E(Y)$, and an increase or decrease is equally likely according to each broker, neither broker's scenario dominates the other.

3.31. Assume that time to fill an order for condensers of all types at a particular company is assumed to have (roughly) a normal distribution. As a result of an influx of new employees, it is claimed that the mean time has increased but not the variance, and the distribution is still approximately normal. Critique the following statement: "The distribution of times may appear to have the same variance, but if the distribution is normal, the variance must have decreased because the increase in time has moved the distribution closer to a conceptual upper boundary on time to fill an order (say, 2 months), so that the range of possible times has decreased and so has the variance."

Solution:
A distinction must be made between a population distribution and an empirical distribution. Although the range of *possible* values has decreased, the variance can still be essentially unchanged if the largest values are well within the upper boundary.

3.33. If X and Y are independent random variables, $\sigma_x^2 = 14$, $\sigma_y^2 = 20$, $E(X) = 20$, and $E(Y) = 40$, what must be the numerical value of $E(XY)$?

Solution:
Since X and Y are independent random variables, $E(XY) = E(X)\,E(Y) = (20)(40) = 800$.

3.35. Assume that $X \sim N(\mu = 50, \sigma^2 = 16)$.

(a) Determine $P(X > 55)$.

(b) Determine $P(X = 48)$.

(c) Determine the lower bound on how large the sample size must have been if $P(\bar{X} > 51) < .06681$.

Solution:
(a) $P(X > 55) = P(Z > 1.25) = .1056$, with $Z = \dfrac{55 - 50}{4}$

(b) zero

(c) The probability .06681 converts to $Z = 1.5$. This is thus the lower bound on

$$Z = \frac{\bar{X} - \mu}{\frac{\sigma}{\sqrt{n}}} = \frac{51 - 50}{\frac{4}{\sqrt{n}}} = \frac{\sqrt{n}}{4} > 1.5 \implies n > 36.$$

3.37. The following question was posed on an Internet message board: "If I have three machines in a department that have the potential to run for 450 minutes per day but on average each one runs for only 64 minutes, what is the probability of them all running at the same time?" The person wanted to know if one of the machines could be replaced, this being the motivation for the question. Can this answer be determined, and will the computed probability address the central issue?

Solution:
More information must be given, especially information about when the machines run (morning, afternoon, etc.). Information must also be known about the distribution of the running time for each machine.

3.39. The term "engineering data" is often used in the field of engineering to represent relationships, such as the following relationship: tensile strength (ksi) = 0.10 + 0.36(CMMH), with CMMH denoting "composite matrix microhardness." If we wanted to model tensile strength in a particular application by using an appropriate statistical distribution, would the relationship between tensile strength and CMMH likely be of any value in selecting a distribution? What information must be known, in general, before one could attempt to select a model?

Solution:

The relationship will not be of any value unless the distribution of CMMH is known.

3.41. Which one of the four moments discussed in the chapter is in the original unit of measurement?

Solution:
The first moment

3.43. If $P(A|B) = P(A)$, show that $P(B|A)$ must equal $P(B)$.

Solution:
$P(A \mid B) = P(A)$ results from the formula for conditional probability only if $P(A \cap B) = P(A)P(B)$. It then follows from this that $P(B \mid A)$ must simplify to $P(B)$ since $P(B \cap A) = P(A \cap B)$.

3.45. Given: $f(x) = 1/4 \quad 0 < x < 4$.

(a) Show that this is a probability density function.

(b) Sketch the graph of the function. Is this one of the distribution types covered in the chapter? Explain.

(c) Determine $Var(X)$.

Solution:
(a) It is a probability function because $\int_0^4 (1/4)dx = 1$
(b) The graph is simply a horizontal line at $Y = f(x) = 1/4$, with the line extending from $X = 0$ to $X = 4$. This is the continuous uniform distribution.
(c) $Var(X) = \int_0^4 x^2(1/4)\, dx - (\int_0^4 x(1/4)\, dx)^2 = 64/12 - 2^2 = 16/12 = 1.33$.

3.47. Given the following *pdf*, determine $E(X)$ and the variance:

X	2	4	6	7
$P(X)$	1/4	1/3	1/6	1/4

Solution:
$E(X) = 2(1/4) + 4(1/3) + 6(1/6) + 7(1/4) = 55/12 = 4.58$
$Var(X) = 4(1/4) + 16(1/3) + 36(1/6) + 49(1/4) - (55/12)^2 = 407/1444 = 0.28$

3.49. Assume that you are given (most of) a *pmf*. If you were given the following, explain why the listed probabilities could not be correct.

X	1	2	3	4	5
$P(X)$	1/8	1/4	1/3	1/2	?

Solution:
The probabilities given could not be correct because they add to more than one.

3.51. Assume that in a very large town 20% of the voters favor a particular candidate for political office. Assume that five voters are selected at random for a television interview and asked for their preferences. If they express their preferences independently, what is the probability that at most one of them will indicate a preference for the other candidate?

Solution:
$X \sim B(5, .2)$ $P(X \geq 4) = P(X = 4) + P(X = 5) = \binom{5}{4}(.2)^4(.8)^1 + \binom{5}{5}(.2)^5(.8)^0$
$= 45/160 = .281$

3.53. Consider a binomial distribution problem with $P(X = n) = 1/128$. What is $P(X = 0)$? Can there be more than one answer? Explain.

Solution:
$P(X = n) = 1/128 = p^n$ so $p = (1/128)^{1/n}$. Since there are two unknowns, there are many possible solutions. One of these is $n = 7$ and $p = 1/2$. Then $P(X = 0) = (1 - p)^n = 1/128$.

3.55. Given the following probability mass function: $f(x) = a$, $2 \leq x \leq 8$, determine the numerical value of a.

Solution:
The function must integrate to 1, so $a = 1/6$.

3.57. Assume that $X \sim N(\mu = 20, \sigma = 3)$. If $Y = 2X - 8$:
   (a) What are the mean and variance of Y?
   (b) Would your answers be different if you weren't asked to assume normality? Explain.

Solution:
   (a) The mean is $2(20) - 8 = 32$ and the variance is $(2)^2(3) = 12$
   (b) Normality or non-normality has no effect on the computations for the mean and variance in part (a).

3.59. Given $f(x) = 4x^3$, $0 < x < a$:
   (a) What must be the value of a for the function to be a *pdf*?
   (b) Explain why $E(X)$ is not midway between 0 and a.

Solution:

(a) $\int_0^a 4x^3 dx = 1$ so $a^4 = 1$ and thus $a = 1$.

(b) The distribution is not symmetric.

3.61. The August 2, 2000 edition of *People's Daily* stated that, according to a national survey of 4,000 people, 91.6% of Chinese are in favor of amending the existing Marriage Law. Approximate the probability that out of 5 randomly selected Chinese from the same population to which the survey was applied, at least 4 will favor an amendment. Why is this an approximation and not an exact value?

Solution:
$\binom{5}{4}(.916)^4(.084) + \binom{5}{5}(.916)^5 = .9406$ This is an approximation because there is a finite population, but n/N is small enough to permit the correction factor to be ignored.

3.63. Acceptance sampling methods are used in industry to make decisions about whether to accept or reject a shipment, for example. Assume that you have a shipment of 25 condensers and you decide to inspect 5, without replacement. You will reject the shipment if at least one of the five is defective.

(a) In the absence of any information regarding past or current quality, does this appear to be a good decision rule?

(b) What is the probability of rejecting the shipment if the latter contains one bad condenser?

Solution:
(a) By today's quality standards, this is a reasonable decision rule.

(b) $\dfrac{\binom{1}{1}\binom{24}{4}}{\binom{25}{5}} = .20$

3.65. Consider the function $f(x) = x^2$ with the range of the function as defined in Exercise 3.64. Is this a *pdf*? Explain.

Solution:
It is not a *pdf* because it integrates to 2/3 instead of 1.

3.67. There is a 25% chance that a particular company will go with contractor A for certain work that it needs to have done, and a 75% chance that it will go with contractor B. Contractor A finishes 90% of his work on time and contractor B finishes 88% of his work on time. The work that the company wanted done was completed on time, but the identity of the contractor was not widely announced. What is the probability that the contractor selected for the job was contractor A?

Solution:
This is a Bayes problem. Using the following probabilities: $P(A) = .25$, $P(B) = .75$, $P(\text{completed} \mid A) = .90$ and $P(\text{completed} \mid B) = .88$, we obtain

$$P(A \mid \text{completed}) = \frac{P(\text{completed} \mid A)\, P(A)}{P(\text{completed} \mid A)\, P(A) + P(\text{completed} \mid B)\, P(B)}$$

$$= \frac{(.90)(.25)}{(.90)(.25) + (.88)(.75)} = .254$$

3.69. Assume that one of the eight members of the customer service department of a particular retailer is to be randomly selected to perform a particular unsavory task. Numbers are to be generated from the unit uniform distribution [i.e., $f(x) = 1$, $0 < x < 1$]. The employees are assigned numbers 1-8 and the first random number is assigned to employee #1, the second random number to employee #2, and so on. The first employee whose assigned random number is greater than 0.5 is assigned the task.

(a) What is the probability that employee #4 is picked for the job?

(b) What is the probability that no employee is selected and the process has to be repeated?

Solution:
(a) $P(\text{first 3 are less than or equal to } .5 \text{ and the 4th is greater than } .5) = (.5)^4 = .0625$.

(b) $P(\text{ none exceed } .5) = (.5)^8 = .0039$

3.71. Bayes' Theorem is often covered in engineering statistics courses. A student is skeptical of its value in engineering, however, and makes the following statement after class one day: "In working a Bayes problem, one is generally going backward in time. I am taking another class that is covering the contributions of W. Edwards Deming. He claimed that 'statistics is prediction', which obviously means that we are moving forward in time. I don't see how any statistical tool that takes us back in time can be of any value whatsoever."

(a) How would you respond to this student?

(b) If possible, give an application of Bayes' theorem from your field of engineering or science. If you are unable to do so, do you agree with the student? Explain.

Solution:
(a) Sometimes in order to go forward we must first go backward. Seriously, in practical applications of Bayes' Theorem, which are often sequential,

(b) (student exercise)

3.73. Assume $X \sim N(\mu, \sigma^2)$. Obtain $E(X^2)$ as a function of μ and σ^2.

Solution:
Since $\sigma^2 = E(X^2) - (E(X))^2 = E(X^2) - \mu^2$, it follows that $E(X^2) = \sigma^2 + \mu^2$.

3.75. Assume that 10 individuals are to be tested for a certain type of contagious illness and a blood test is to be performed, initially for the entire group. That is, the blood samples will be combined and tested, as it is suspected that no one has the illness, so there will be a cost saving by performing a combined test. If that test is positive, then each individual will be tested. If the probability that a person has an illness is .006 for each of the 10 individuals, can the probability that the test shows a positive result be determined from the information given? If not, what additional information is needed?

Solution:
The probability cannot be determined from the information that is given. There are two important pieces of information that are missing: (1) what is the probability of the test showing positive when illness is present, and (2) since this is a contagious disease, have any of the 10 people who are being tested previously been in contact with at least one member of the group?

3.77. Becoming proficient at statistical thinking is more important than the mastery of any particular statistical technique. Consider the following. There are certain organizations that have limited their memberships to males and have not had female members. A person applies for membership with his or her height indicated, but gender not filled in. Since the name "Kim" can signify either a man or a woman, the membership committee is unsure of the gender of an applicant with this first name. The person's height is given as 5 ft 7 1/2 in. The mean heights of all men and all women are of course known (approximately), and assume that the standard deviations are also known approximately. Assume that heights are approximately normally distributed for each population and explain how the methods of this chapter might be used to determine whether the applicant is more likely a woman or a man.

Solution:
Two sets of Z-scores might be computed, with each pair obtained so as to give $P(67$ inches $\leq$ height ≤ 68 inches) for men and for women, with the choice determined from the larger of the two probabilities.

3.79. Service calls come to a maintenance center in accordance with a Poisson process at a rate of $\lambda = 2.5$ calls every 5 minutes. What is the probability that at least one call comes in 1 minute?

Solution:
The Poisson mean for one minute is 0.5. Let X = number of calls. $P(X \geq 1)$ =
$1 - P(X = 0) = 1 - .6065 = .3935$.

3.81. The lifetime of a device has an exponential distribution with a mean of 100
hours. What is the probability that the device fails before 100 hours have passed?
Explain why this probability differs considerably from .5 even though the mean is
100.

Solution:
$f(x) = \frac{1}{100}e^{-x/100}$ $P(X < 100) = \int_0^{100} \frac{1}{100}e^{-x/100}\,dx = 1 - e^{-1} = .632$. This
differs considerably from .50 because the exponential distribution is a right-
skewed distribution, so most of the probability is at the lower end.

3.83. Assume that X has an exponential distribution with a mean of 80. Determine
the median of the distribution.

Solution:
The median is 55.4518. This is most easily obtained by using appropriate software,
as the following MINITAB commands will produce the answer: MTB> INVCDF
80; SUBC>EXPO 80. Alternatively, the following integration could be performed:
$\int_{-\infty}^{c} (1/80)exp(-x/80) = .50$ and solve for c.

3.85. Assume that all units of a particular item are inspected as they are
manufactured. What is the probability that the first nonconforming unit occurs on
the 12th item inspected if the probability that each unit is nonconforming is .01?

Solution:
Using the geometric distribution, the solution is $(1 - .01)^{99}(.01) = .009$.

3.87. Consider a particular experiment with two outcomes for which the binomial
distribution seemed to be an appropriate model, with $p = 1/8$. The probability of
$n - 1$ successes is equal to 350 times the probability of n successes. What is the
numerical value of n?

Solution:
$P(n - 1 \text{ successes}) = 350\, P(n) = 350 \left(\frac{1}{8}\right)^n = n\left(\frac{1}{8}\right)^{n-1}\left(\frac{7}{8}\right) \Rightarrow 7n = 350$ so $n = 50$.

3.89. (Harder problem) Two students in your engineering statistics class decide
to test their knowledge of probability, which has very recently been covered in
class. They each use MINITAB or other statistical software to generate a number
that has a continuous uniform distribution over a specified interval. The first
student uses the interval 0-3 and the second student uses the interval 0-1. Before
the random numbers are obtained, the students compute the probability that the
first student's number will be at least twice the second student's number.

(a) What is that probability?

(b) What is the probability after the random numbers have been obtained? Explain.

(c) What is the probability that the maximum of the two numbers is at most 1/2?

(d) What is the probability that the minimum of the two numbers is at least 1/4?

Solution:

(a) The space for the experiment is a rectangle formed by the vertical lines $x = 0$ and $x = 1$ and the horizontal lines $y = 0$ and $y = 3$. The appropriate distribution (not covered in the text) is a bivariate continuous distribution. The problem may be worked without using that distribution, however. All regions of the same size within the rectangle have equal probability, so we simply need to determine the area below the line $y = 2x$. Since this is a "left triangle", the area is $(1/2)(1)(2) = 1$. Thus the area above the line is $3 - 1 = 2$, so the probability is 2/3.

(b) The probability is, trivially, either zero or one depending on whether the first student's number was bigger or not (i.e, we cannot compute a non-trivial probability after the fact).

(c) This is the probability that both numbers are less than 1/2. This comprises an area of 1/4, so the probability is 1/12.

(d) This probability is the relative area of the rectangle formed by the vertical lines $x = 1/4$ and $x = 1$ and the horizontal lines $y = 1/4$ and $y = 3$. That area is $(3/4)((11/4) = 33/16$ and $(33/16)/3 = 11/16$.

3.91. You notice that only 1 out of 20 cars that are parked in a particular tow-away zone for at least 10 minutes is actually towed. Because of very bad weather, you are motivated to park your car as close as possible to a building in which you have short, 10-15-minute meetings for two consecutive days. You decide to take a chance and park in the tow-away zone for your meetings on those two days. You decide, however, that if your car is towed on the first day that you won't park there on the second day. What is the probability that your car will be towed? (Have you made an assumption in working the problem that might be unrealistic? Explain.)

Solution:
$P(\text{towed}) = P(\text{towed on first day}) + P(\text{towed on second day}|$ not towed on first day$) = .05 + .05 = .10$. This is based on the assumption of independent events, which might not be a valid assumption as the police might be more likely to have a car towed if they see it parked in the tow-away zone on the second day after seeing the same car leave the tow-away zone on the previous day.

3.93. Two students in a statistics class are told that their standard scores (i.e., z-scores as in Section 3.4.3 except that $\bar{x}$ and s are used instead of μ and σ, respectively) on a test were 0.8 and -0.2, respectively. If their grades were 88 and 68, what were the numerical values of $\bar{x}$ and s?

Solution:
The information that is given produces two equations in the two unknowns, $\bar{x}$ and s, with each equation of the form $Z = (x - \bar{x})/s$. Solving for the two unknowns produces $\bar{x} = 72$ and $s = 20$.

3.95. Consider the probability statement $P(2Z_{\alpha/2} > z_0) = .236$, with $Z \sim N(0, 1)$. Determine the numerical value of z_0. What is the numerical value of α?

Solution:
$P(2Z_{\alpha/2} > z_0) = .236 \Rightarrow P(Z_{\alpha/2} > z_0/2) = .236$. Using software, we obtain $P(Z_{\alpha/2} > 0.7192) = .236$ so $z_0 = 2(.7192) = 1.4384$. Since $\alpha/2 = .236$, $\alpha = 2(.236) = .472$.

3.97. A department with seven people is to have a meeting and they will all be seated at a round table with seven chairs. Two members of the department recently had a disagreement, so it is desirable that they not be seated next to each other. If this requirement is met but the seating is otherwise assigned randomly, how many possible seating arrangements are there and what is the probability associated with any one of those arrangements?

Solution:
With no restrictions, there are 7! possible arrangements. There are 5! arrangements of the other 5 people for a given seating combination of the two people who can't sit next to each other. There are 7 seats in which they could be seated with one on the right of the other and 7 with one on the left of the other. So the number of seating arrangements that would be acceptable is thus 7! - 7(2)(5!) = 3360. The probability associated with one of these arrangements is 1/3360.

3.99. Given that $P(X = 1) = 2P(X = 4)$, complete the following table and determine the expected value of X.

x	1	4	6
$P(x)$	_	_	1/6

Solution:
$2(P(X = 4)) + P(X = 4) = 5/6$, so $P(X = 4) = 5/18$ and $P(X = 1) = 10/18$. Then $E(X) = 1(10/18) + 4(5/18) + 6(1/6) = 48/18 = 2.67$.

4

Point Estimation

4.1. Show that the mean and variance of $\overline{X}$ are μ and σ^2/n, respectively, with μ and σ^2 denoting the mean and variance, respectively, of X. Does the result involving the mean depend on whether or not the observations in the sample are independent? Explain. Is this also true of the variance? Is there a distributional result that must be met for these results to hold, or is the result independent of the relevant probability distribution?

Solution:

$E(\overline{X}) = E[(X_1 + X_2 + \ldots X_n)/n] = (1/n)E(X_1 + X_2 + \ldots X_n) =$
$(1/n)\sum_{i=1}^{n}E(X_i) = (1/n)\sum_{i=1}^{n}\mu = (1/n)(n\mu) = \mu.$
$Var(\overline{X}) = Var[(X_1 + X_2 + \ldots X_n)/n] = (1/n^2)Var(X_1 + X_2 + \ldots X_n) =$
$(1/n^2)\sum_{i=1}^{n}Var(X_i) \text{ (if the } X_i \text{ are independent)} = (1/n^2)(n\sigma^2) = \sigma^2/n$

The result for the mean does not require independence, but the variance result does require independence. An assumed distribution isn't necessary for either result.

4.3. Critique the following statement: "I see an expression for a maximum likelihood estimator for a particular parameter of a distribution that I believe will serve as an adequate model in an application I am studying, but I don't see why a scientist should be concerned with such expressions as I assume that statistical software can be used to obtain the point estimates."

Solution:
Maximum likelihood estimates are generally available in statistical software.

4.5. Assume that $X \sim N(\mu = 20, \sigma^2 = 10)$ and consider the following two estimators of μ, each computed from the same random sample of size 10.

$\hat{\mu}_1 = \sum_{i=1}^{10} c_i x_i$ with $x_1, x_2, ... x_{10}$ denoting the observations in the sample, and

$$c_i = 0.2, \quad i = 1, 2, ..., 5$$
$$= 0, \quad i = 6, 7, ...,10$$

$\hat{\mu}_2 = \sum_{i=1}^{10} k_i x_i$ with the x_i as previously defined and

$$k_i = 0.10 \text{ for } i = 1, 2, ..., 9$$
$$= 0 \text{ for } i = 10.$$

(a) Determine the variance of each estimator.

(b) Is each estimator biased or unbiased? Explain.

(c) Is there a better estimator? If so, give the estimator. If not, explain why a better estimator cannot exist.

Solution:
(a) $Var(\hat{\mu}_1) = 5(0.2)^2(10) = 2.0$
$\quad Var(\hat{\mu}_2) = 9(0.1)^2(10) = 0.9$

(b) The first estimator is unbiased because the weights sum to 1.0; the second estimator is biased because the weights do not sum to 1.0.

(c) The best estimator ("best" in terms of having the smallest variance among unbiased estimators) is the sample mean, $\bar{x}$.

4.7. Assume that $X \sim N(10, \sigma^2)$.

(a) Obtain the maximum likelihood estimator for σ^2.

(b) Is the estimator unbiased? Explain. If biased, what must be done to convert it to an unbiased estimator?

Solution:
(a) The joint *pdf* is $(2\pi)^{-n/2} \sigma^{-n} \exp[-1/2 \sum[(x - 10)/\sigma]^2$, the *log* of which is

$-1/2\sigma^2 \sum(x - 10)^2 - log(\sigma^n) - $ constant.   Taking the derivative of this and

setting it equal to zero , we obtain $\sum(x - 10)^2/\sigma^3 - n/\sigma = 0$. Solving for σ^2

we thus obtain $\hat{\sigma}^2 = \sum(x - 10)^2/n$.

(b) The estimator is biased; it would have to be multiplied by $(n/(n - 1))$ to make it unbiased.

4.9. Assume that a sample of 10 observations has been obtained from a population whose distribution is given by $f(x) = 1/5$, $x = 1, 2, 3, 4, 5$. If possible, obtain the maximum likelihood estimator of the mean of this distribution. If it is not possible, explain why it isn't possible.

Solution:
It is not possible to obtain the maximum likelihood estimator of the mean because the mean is not a parameter in the *pdf*.

4.11. Assume that a random variable has a distribution with considerable right skewness (i.e., the tail of the distribution is on the right). Describe as best you can what the distribution of the sample mean will look like relative to the distribution of X.

Solution:
The distribution of the sample mean will also exhibit right skewness, but will have a smaller variance.

4.13. Assume that a random variable has a distribution that is approximately normal, and also assume that it would be rare for an observation from the population that has this distribution to be either larger than 120 or smaller than 30. Using only this information, what would be your estimate of σ^2?

Solution:
Since the population is approximately normally distributed, the standard deviation would be estimated as the range divided by 6 (based on $\mu \pm 6\sigma$ covering almost the entire range of possible values). Thus, for this example σ would be estimated as $(120 - 30)/6 = 15$, so the estimate of σ^2 is $(15)^2 = 225$.

4.15. Given $f(x) = 2x$, $0 < x < 1$, determine the method of moments estimator of μ if possible. If it isn't possible, explain why.

Solution:
The mean, which is 2/3, is known because it is not a function of any unknown parameters. Therefore, we cannot speak of a method of moments estimator for something that does not have to be estimated.

4.17. Assume that $X \sim N(50, 25)$. If 1,000 samples are produced in a simulation exercise and the range of the $\overline{X}$ values is 47.3 to 52.8 but the sample size was not recorded, what would be your "estimate" of the sample size?

Solution:
Since X is normally distributed, $\overline{X}$ is also normally distributed. With 1,000 samples generated, the largest and smallest $\overline{X}$ values should be in the neighborhood of

$3\sigma_{\bar{X}}$ from μ. Thus, $3 = \frac{52.8-50}{\frac{5}{\sqrt{n}}} \Rightarrow n = 29$ so 29 would be a reasonable "estimate" of n.

4.19. Explain the difference between an estimate and an estimator.

Solution:
An estimator is a random variable; it has a mean, variance and distribution. An estimate is a number in the case of a point estimate and an interval in the case of an interval estimate. Estimates do not have the properties that estimators do because estimates are simply numbers.

4.21. Consider the standard normal distribution in Section 3.4.3. Obtain a random sample of 25 from that distribution, using MINITAB or other software. We know that $\sigma_{\bar{x}} = 0.2$. Obtain 1,000 (bootstrap) samples of size 10 and estimate $\sigma_{\bar{x}}$. (It will be necessary to write a macro in MINITAB to accomplish this.) Comment on your results relative to the known value.

Solution:
(simulation exercise performed by reader)

4.23. Can the maximum likelihood estimator of β for the exponential distribution given in Section 3.4.5.2 be obtained using the approach illustrated in Section 4.4.1? If so, obtain the estimator. If not, explain why not and determine if the estimator can otherwise be determined.

Solution:
Yes, the maximum likelihood estimator can be obtained as follows. With $f(x) = \frac{1}{\beta}e^{-x/\beta}$, the likelihood function is $L = \frac{1}{\beta^n}e^{-\sum x/\beta}$. Thus, $log(L) = -n\,log(\beta) - \sum x/\beta$ so $\frac{d\,log(L)}{d\beta} = -\frac{n}{\beta} + \frac{\sum x}{\beta^2} = 0 \rightarrow -n\beta + \sum x = 0$ and $\hat{\beta} = \frac{\sum x}{n}$ is the maximum likelihood estimator.

4.25. Is it possible to obtain the method of moments estimators for the mean and variance of the continuous uniform distribution (see Section 3.4.2) defined on the interval $(0,1)$? If so, obtain the estimators; if not, explain why it isn't possible.

Solution:
Method of moments estimators are obtained by equating population moments to sample moments. The continuous uniform distribution defined on a specific interval does not contain any parameters, however, so equating sample moments does not lead directly to estimators in a general form. Furthermore, the mean and variance are 1/2 and 1/12, respectively, so there is no need to estimate them.

(Equating the sample mean to the population mean would produce the nonsensical result $\bar{x} = 1/2$.)

4.27. A store decides to conduct a survey regarding the distribution of men and women that visit the store on Saturdays. A particular Saturday is chosen and an employee is told to record a "1" whenever a man enters the store and a "2" whenever a woman enters. Let X denote the number of women who enter the store. Can $Var(X)$ be determined based on this information? If so, what is the variance? If not, explain why the variance cannot be obtained.

Solution:
The variance cannot be determined without knowing the total number of people who entered the store since the variance is a function of that number.

4.29. Show that S^2 is an unbiased estimator of σ^2 when a normal distribution is assumed.

Solution:
From Section 3.4.5.1 we know that $\sum_{i=1}^{n}(\frac{X_i-\mu}{\sigma})^2$ is χ_n^2 when $X_i \sim N(\mu, \sigma^2)$. We also know from that section, at least indirectly, that the distribution is χ_{n-1}^2 when μ is replaced by $\bar{x}$. Since $\sum_{i=1}^{n}(\frac{X_i-\mu}{\sigma})^2 = (1/\sigma^2)\sum_{i=1}^{n}(X_i - \mu)^2$ and replacing μ by $\bar{x}$ gives $(1/\sigma^2)\sum_{i=1}^{n}(X_i - \bar{x})^2 = (1/\sigma^2)[(n-1)s^2] \sim \chi_{n-1}^2$. It was also stated in Section 3.4.5.1 that the mean of a chi-square random variable is the degrees of freedom, so $E(s^2) = (\sigma^2/(n-1))E(\chi_{n-1}^2) = (\sigma^2/(n-1))(n-1) = \sigma^2$. Thus, s^2 is an unbiased estimator of σ^2.

4.31. Show that $\sum_{i=1}^{n}(x_i - a)^2$ is minimized when $a = \bar{x}$. Does this mean that $\bar{x}$ is thus the least squares estimator of μ_x? Explain.

Solution:
Let $a = \bar{x} + \epsilon$. Then $\sum_{i=1}^{n}(x_i - a)^2 = \sum_{i=1}^{n}(x_i - (\bar{x} + \epsilon))^2 = \sum_{i=1}^{n}(x_i - \bar{x}) - \epsilon)^2$
$= \sum_{i=1}^{n}(x_i - \bar{x})^2 + n\epsilon^2 > \sum_{i=1}^{n}(x_i - \bar{x})^2$ for any real number ϵ. (The middle term vanishes since $-2\epsilon\sum_{i=1}^{n}(x_i - \bar{x}) = 0$ since $\sum_{i=1}^{n}(x_i - \bar{x}) = 0$.)

4.33. Consider the information given in Exercise 1.63.

(a) What must be assumed in order to estimate σ for the fraternity GPA using any method given in this chapter? Does such an assumption seem plausible? If so, what would be the estimate?   If the assumption seems implausible, explain why.

(b) Could the same approach be used to estimate the standard deviation of the sorority GPA? Explain.

Solution:
(a) In the absence of the raw data, it would be necessary to assume approximate normality in order to use the range method of estimating sigma. It was stated that the distribution of the fraternity GPAs is close to symmetric, so we might assume approximate normality. We don't have the extreme values, but we do have the largest value and the third quartile. Under normality, $\mu + 0.6745\sigma = 2.7535$. Letting $\mu + 3\sigma$ correspond to the largest value of 3.130, we can solve these two equations and obtain an estimate of σ as $\hat{\sigma} = 0.16$.

(b) No, the distribution of sorority GPAs is clearly skewed, as evidenced by the fact that the distance from the smallest value to Q_1 is 0.301, whereas the distance from Q_3 to the largest value is only 0.074. Therefore the same approach could not be used and there is nothing that can be done in the absence of the raw data or information about the distribution.

4.35. Explain why the maximum likelihood estimator of the mean of a normal distribution has the same variance as the method of moments estimator.

Solution:
The variances are the same because the estimators are the same

4.37. A manager is considering giving employees in her division a 4% cost-of-living raise plus a flat $600 bonus in addition to the raise. Assume that the employees in her division currently have a mean salary of $43,000 with a standard deviation of $2,000. What will be the new mean and the new standard deviation after both the raise and the bonus?

Solution:
The new mean is $(1.04)(43,000) + 600 = \$45,320$. The new standard deviation is $(1.04)(2,000) = \$2,080$.

4.39. Explain why there is no maximum likelihood estimator for the *pdf* $f(x) = 1$, $0 < x < 1$.

Solution:
There is no maximum likelihood estimator for this *pdf* because there is no parameter to estimate in the *pdf*.

4.41. Assume that you are a supermarket manager and you want to obtain some idea of the variability in the amount of time required to check out a customer in the express lane once the customer reaches the front of the line compared to the amount of time for customers in the lanes that do not have a limit on the number of items. You know from past data that customers in the other lanes have an average of 29.8 items, whereas customers in the express lane have an average of 8.9 items. If the estimate of the standard deviation for the other lanes is 2.2 minutes, what would be the estimate of the standard deviation for the express lane, or is it possible to even obtain an estimate from this information? Explain.

Solution:
It would be reasonable to assume that the coefficients of variation for the express lanes and other lanes do not differ greatly. Under this assumption, a ballpark estimate for the standard deviation for the express lane is 0.657, as this number equates the coefficients of variation.

4.43. Data in a population database that are in inches are converted to centimeters. What is the variance of the new data as a function of the variance of the old data?

Solution:
Let the variance for the original data, in inches, be given by c. Since measurement in centimeters = $2.54 \times$ measurement in inches, the variance in centimeters as a function of c is $(2.54)^2 c = 6.45c$.

4.45. Information given at the Energy Information Administration website (www.eia.doe.gov) showed that in 2000 the interquartile range for underground mines in Alabama was $22.54 per short ton. Assuming a normal distribution, could σ be estimated from this information? If so, what would be the estimate? If not, explain why an estimate cannot be obtained.

Solution:
Yes, under the assumption of normality, the interquartile range is $\mu + 0.6745\sigma - (\mu - 0.6745\sigma) = 1.3490\sigma$. So $1.3490\sigma = \$22.54$ and $\hat{\sigma} = \$16.71$.

5

Confidence Intervals and Hypothesis Tests -- One Sample

5.1. Show that the form of a confidence interval for μ with σ estimated from a small sample is as was given in Section 5.2 by using the same general starting point as was used in Section 5.1 and then deriving the expression for the interval.

Solution:
The starting point is the following probability statement, analogous to the starting point used in Section 5.1:

$$P(-t_{\alpha/2,\,n-1} \le \frac{\overline{X}-\mu}{s/\sqrt{n}} \le t_{\alpha/2,n-1}) = 1 - \alpha$$

Multiplying through by $s/\sqrt{n}$, subtracting $\overline{X}$, and then multiplying through by -1 produces

$$P(\overline{X} - t_{\alpha/2,\,n-1}\,s/\sqrt{n} \le \mu \le \overline{X} + t_{\alpha/2,\,n-1}\,s/\sqrt{n}) = 1 - \alpha$$

with the endpoints giving the $100(1 - \alpha)\%$ confidence interval for μ.

5.3. Show that if a two-sided hypothesis test for $H_0: \mu = \mu_0$ with significance level α and normality and an assumed known σ is *not* rejected, then the corresponding $100(1 - \alpha)\%$ confidence interval *must* contain μ_0.

Solution:
Since the null hypothesis is not rejected, $Z = \frac{\overline{X}-\mu_0}{\sigma/\sqrt{n}}$ is between $-Z_{\alpha/2}$ and $Z_{\alpha/2}$. Since this is the starting point for producing the form of the confidence interval, it follows that μ_0 must lie inside the interval.

5.5. Assume that a sample of 64 observations will be obtained from a population that has a normal distribution. Even though σ is unknown, the practitioner decides to obtain a 95% confidence interval for μ using Z (i.e., using 1.96). What is the actual probability that the interval will contain μ? (*Note:* You may find it necessary to use appropriate software in working this problem.)

Solution:
$P(t_{63} > 1.96) = .0272 \Rightarrow$ degree of confidence is $100(1 - .0272)\% = 94.56\%$

5.7. One of my professors believed that students should take four times as long to finish one of his exams as it took him to complete it. He completes a particular exam in 15 minutes, so he reasons that the students should complete the exam in 60 minutes. (The students are allowed 70 minutes.) Assume that each of 40 students in a class writes down the length of time that it took to complete the first test. The times are as follows, to the nearest minute: 55, 61, 58, 64, 66, 51, 53, 59, 49, 60, 48, 52, 54, 53, 46, 64, 57, 68, 49, 52, 69, 60, 70, 51 50, 64, 58, 44, 63, 62, 59, 53, 48, 60, 66, 63, 45, 64, 67, and 58. A student complains that the test was too long and didn't provide her sufficient time to look over her work. (Assume that 5 minutes should be sufficient for doing so, and this time was now included in the professor's estimate.) Perform the hypothesis test and state your conclusion.

Solution:
We wish to test $H_0: \mu = 60$ vs. $H_a: \mu > 60$. We obtain the following. Since $\overline{X} = 57.33$ is less than 60, the null hypothesis would not be rejected. Thus, there is no evidence that the allotted time should not have been sufficient, and a 95% upper confidence bound is $57.33 + 1.6849 \frac{7.15}{\sqrt{40}} = 59.2$, which does not cover 60.

5.9. An experimenter constructs a 95% two-sided confidence interval for μ, using a sample size of 16. Normality of the individual observations is assumed and the population standard deviation is unknown. The limits are: lower limit = 14, upper limit = 24. If instead of constructing the confidence interval, the experimenter had tested $H_0: \mu = 10$ against $H_a: \mu \neq 10$ using $\alpha = .05$, what was the numerical value of the test statistic?

Solution:
The value of the sample standard deviation, s, was not given, but can be solved for using the two expressions that produce the two limits. That is,

$$\overline{x} + t_{.025, 15} \, s/\sqrt{n} = 24$$
$$\overline{x} - t_{.025, 15} \, s/\sqrt{n} = 14$$

Subtracting the second equation from the first one and substituting $t_{.025, 15} = 2.131$ produces $2(2.131)s/\sqrt{16} = 10 \Rightarrow s = 9.38$. The value of $\overline{x}$ must be 19 since $\overline{x}$ must be in the middle of the confidence interval. The value of the test statistic is then

$$t = \frac{19 - 10}{\frac{9.38}{\sqrt{16}}} = 1.92$$

5.11. Given a particular data set, would a null hypothesis for a two-sided test that is rejected by one data analyst using $\alpha = .01$ also be rejected by another analyst using $\alpha = .05$ if each person used the same test procedure so that the only difference was α? (Assume that no errors are made.) Now assume that the analyst who used $\alpha = .01$ did *not* reject the null hypothesis. Can the outcome be determined for the other analyst who used $\alpha = .05$? Explain.

Solution:
If a null hypothesis is rejected using $\alpha = .01$, then it would have to be rejected using $\alpha = .05$, since the former is the stronger test. If the null hypothesis was not rejected using $\alpha = .01$, then we don't know whether or not it could have been rejected using $\alpha = .05$.

5.13. Assume that $H_0: \mu = 25$ is rejected in favor of $H_a: \mu \neq 25$ using $\alpha = .01$. Assume further that a 99% two-sided confidence interval is subsequently constructed using the same set of data that was used to test the hypothesis. If the value of the appropriate test statistic was negative, the upper limit of the confidence interval must have been _____. (Use either "greater than" or "less than" as part of your answer.)

Solution:
"Less than 25" because both endpoints of the interval must be less than the hypothesized value.

5.15. Assume that your driving time to school/work is (approximately) normally distributed and you wish to construct a confidence interval for your (long-term) average driving time. So you record the time for 10 consecutive days and find that the average is 23.4 minutes and the standard deviation is 1.2 minutes. If you wish to construct a 99% confidence interval for what your average driving time would be for a very long time period, what assumption will you have to make? With that assumption, what is the interval?

Solution:
You have to assume that the driving conditions during those 10 days are typical of the driving conditions that you will face in the future. The interval is obtained as $23.4 \pm 3.25(1.2/\sqrt{10})$, with $3.25 = t_{.005,9}$. The interval is thus (22.17, 24.63).

5.17. An experimenter performs a hypothesis test of $H_0: \mu = \mu_0$ versus $H_a: \mu > \mu_0$, using a sample of size 100, and obtains a p-value of .05. Since this is a borderline value, the test is repeated and the absolute value of the test statistic is greater than with the first test. Knowing only this information, what course of action would you take/suggest?

Solution:
I would not recommend that any action be taken based on the stated result of the second test because the value of the test statistic could have been negative, in which case the null hypothesis would not be rejected.

5.19. A process engineer decides to estimate the percentage of units with unacceptable flatness measurements. A sample of size 100 is obtained and a 95% confidence interval is obtained using the standard approach: $\hat{p} \pm$

$Z_{\alpha/2}\sqrt{\hat{p}(1-\hat{p})/n}$. The computed lower limit is negative, which causes the engineer to proclaim that constructing such an interval is a waste of time since a proportion cannot be negative. Explain to the engineer why the lower limit is negative and suggest a superior approach, if one is available.

Solution:
The computed lower limit is negative because the percentage of unacceptable flatness measurements in the sample was small. This type of problem can be avoided by using a better method of constructing the interval, such as the one given in Equation (5.4).

5.21. Explain why we cannot state, for example, that "we are 95% confident that μ is between 45.2 and 67.8."

Solution:
The population mean μ is either between the two numbers or it isn't. In order to make a probability statement, a random variable must be involved, whereas the two endpoints of the intervals are numbers and μ is also a number.

5.23. An experimenter decides to construct a 95% confidence interval for μ and starts to use t because σ is unknown. Someone else in her department informs her that she could use Z because the sample size is $n = 36$. What will be the numerical difference between the widths of the two confidence intervals if $s = 12.1$?

Solution:
Assuming that approximate normality exists so that it is appropriate to use either t or Z, the width of the interval is $2(t \text{ or } Z)s/\sqrt{n}$. If t were used, the width would be $2(2.03)(12.1)/\sqrt{36} = 8.19$. If Z were used, the width would be $2(1.96)(12.1)/\sqrt{36} = 7.91$. Thus, the widths differ by 0.28.

5.25. A 99% confidence interval for p is desired that has width 0.04. Can such an interval be constructed? Explain.

Solution:
The width of a confidence interval for p depends upon $\hat{p}$, the sample proportion. Since there won't be a value for $\hat{p}$ until a sample is obtained, it is not possible to construct a confidence interval for p that will have the desired width.

5.27. Assume that a 95% confidence interval is to be constructed for μ. If $n = 41$, $s = 10$, and normality is assumed, we know that we could use either Z (as an approximation) or t in constructing the interval. If Z is used and the width of the confidence interval is a, what would be the width of the interval, as a multiple of a, if t had been used instead of Z?

Solution:

The width of the interval using Z is $2(1.96)\frac{10}{\sqrt{41}} = a$. The appropriate t-value is $t_{.025, 40} = 2.02$. Since $2.02/1.96 = 1.03$, the interval using t would have a width of $1.03 a$.

5.29. I once had a student who could not understand the concept of a p-value, despite several different ways in which I tried to explain it. Consider one of your relatives who has never taken a statistics course (assuming that you have such a relative), and try to give a more intuitive definition than was given in Section 5.8.

Solution:
"We take samples for the purpose of computing statistics that are used for description and for making inferences about population quantities. If a sample statistic is extreme (large or small) relative to a hypothesized value of the corresponding population quantity, the probability of observing a value for the sample statistic that is at least as extreme as was observed (with "extreme" depending upon whether a larger or smaller value is indicated in an alternative hypothesis, or whether one is testing against "inequality", with the direction of the latter not specified). That probability is termed the p-value."

5.31. The planet Saturn has a relative mass that is 95 times the relative mass of the planet Earth. Does this mean that the difference in relative masses is statistically significant? In general, what does "statistically significant" mean in the context of hypothesis testing?

Solution:
"Statistically significant" means that use of the appropriate sample statistic(s) leads to rejection of the null hypothesis. Comparing the relative mass of Saturn with the relative mass of Earth is unrelated to hypothesis testing.

5.33. If 3000 samples of $n = 30$ and the same number of samples of $n = 99$ are generated from a particular right-skewed distribution, what relationship should be observed for:

(a) The distribution of the sample averages for the two sample sizes.

(b) The *average* of the 3000 sample averages for the two sample sizes.

Solution:
(a) The distribution should be less-skewed for the larger sample size.

(b) The relationship between the averages is not affected by the sample sizes. This is because the $E(\overline{X})$ is a direct function of $E(X)$, with the latter being the mean of the population distribution, which is independent of any sample size.

5.35. Consider Example 5.1 in Section 5.1.2. Explain why a normal distribution is not automatically ruled out as a possible model for the amount of pressure required to remove the bottle top just because no one will be exerting a negative amount of pressure to remove the top, but negative values are included in the range of possible values for any normal distribution.

Solution:
In practice, not very many random variables can have negative values. A normal distribution can still be used as a model, however, because a normal distribution whose variance is small compared to the mean would have only a very tiny fraction of its area below zero, so the fact that the range of possible values for any normal distribution is from minus infinity to plus infinity is not necessarily relevant.

5.37. An important aspect of the use of statistical methods is selecting the appropriate tool for a stated objective. Assume that a company has to meet federal requirements regarding the stated weight of its breakfast cereal. If that weight is 24 ounces, why would we not want to take a sample and test H_0: $\mu = 24$? Instead of this, what would you suggest that the company do in terms of methodology presented in this chapter?

Solution:
Information about variability of individual observations relative to prescribed limits is not contained in the confidence interval for a mean. A tolerance interval is one approach that could be used, and a confidence interval on the proportion of observations above a specified value (such as 23.5 ounces in this case), could also be useful.

5.39. Assume that the width of a 95% small-sample confidence interval for μ is 6.2 for some sample size. Several months later the same population is sampled and another 95% confidence interval for μ is constructed, using the same sample size, and the width is found to be 4.3. What is the ratio of the earlier sample variance to the more recent sample variance?

Solution:
The ratio of the widths is the same as the ratio of the sample standard deviations since the width, w, is given by $w = 2\, t_{\alpha/2, n-1}\, s/\sqrt{n}$. So the ratio is 6.2/4.3 = 1.44. The ratio of the sample variances is then the square of this number, which is 2.08.

5.41. Assume that a sample of $n = 100$ observations was obtained, with $\bar{x} = 25$ and $s = 10$. Based solely on this information, would it be practical to construct a 90% confidence interval for μ? Why or why not?

Solution:
Because of the large sample size, it would be reasonable to construct a confidence interval using Z, as only for some extreme distributions would a sample size of 100 be inadequate.

5.43. Two practitioners each decide to take a sample and construct a confidence interval for μ, using the t-distribution. If one constructs a 95% interval and the other constructs a 99% interval, explain how the widths of the two intervals could be the same.

Solution:
Assuming the same sample size, the practitioner who constructed the 99% confidence interval might have taken a sample for which the value of s was much smaller than the value of s in the sample taken by the other practitioner. It is possible (although, of course, unlikely) that the difference in the degree of confidence could be exactly offset by the difference in the s values so that the width would be the same. Of course if the sample sizes differed and the practitioner who constructed the 99% interval took a much larger sample, the intervals could have the same width if the s values differed very little, or were even the same.

5.45. Assume that we computer-generate 10,000 lower 95% confidence bounds for p.

(a) What would be the best guess of the number of bounds that do exceed p?

(b) Would the expected number of bounds that exceed p be more likely to equal the actual number if only 1,000 intervals had been constructed? Why or why not?

Solution:
(a) $10,000\,(.95) = 9500$

(b) The number of intervals that exceed the lower bound is itself a binomial random variable. Let X denote that random variable, with $X_{1,000}$ and $X_{10,000}$ denoting the number for 1,000 and 10,000 randomly generated samples, respectively. $E(X_{1,000}) = 1,000(.95) = 950$ and $E(X_{10,000}) = 10,000(.95) = 9500$. The probability that a binomial random variable is equal to its mean decreases as the sample size increases. Here $P(X = 950 \mid n = 1000$ and $p = .95) = .0578$, whereas $P(X = 9500 \mid n = 10000$ and $p = .95) = .0183$.

5.47. If $H_0 : \mu = 40$ was rejected in favor of $H_a : \mu < 40$ with the random variable X assumed to have (approximately) a normal distribution, $n = 16$, $s = 5$, and $\alpha = .05$, the value of the test statistic (t or Z) must have been less than _____.

Solution:
1.753 (obtained using either the t-table or appropriate software).

5.49. The following explanation of a confidence interval is given on the Web (http://onlineethics.org/edu/ncases/EE18.html): "If we performed a very large number of tests, 95% of the outcomes would lie in the indicated 95% bounded range." Do you consider this to be an acceptable explanation of a confidence interval? Explain. If not, how would you modify the wording?

Solution:
The explanation is vague and misleading. If we construct a large number of confidence intervals, approximately 95% of them will contain the parameter (or a function of multiple parameters) for which the interval was constructed.

5.51. Assume that a 99% confidence interval is constructed for μ with an assumed known value of $\sigma = 2$. What is the variance of the width of the interval?

Solution:
For a constant n, the variance is zero because σ is assumed known.

5.53. State three hypotheses that would be useful to test in your area of engineering or science and, if possible, test one of those hypotheses by taking a sample and performing the test.

Solution:
(student exercise)

5.55. A company report shows a 95% confidence interval for p with the limits not equidistant from $\hat{p}$. A manager objects to the numbers, claiming that (at least) one of them must be wrong. Do you agree? Explain. If not, what would be your explanation to the manager?

Solution:
Especially when $\hat{p}$ is small, the limits should be constructed using Equations (5.4), which will result in the limits not being equidistant from $\hat{p}$. The latter is not a shortcoming, however, and it is better to have limits that are not equidistant from $\hat{p}$ than to not have a lower limit.

5.57. As stated in Section 5.1, a confidence interval can be constructed to have a maximum error of estimation with a stated probability, with this maximum error of estimation being the half-width of the interval.

(a) Explain, however, why a confidence interval for a normal mean can be constructed to have a specified half-width without the use of software or trial-and-error only if σ is known.

(b) Explain how the expected half-width would be determined when σ is unknown and $n=16$.

Solution:
(a) If σ is unknown, there are two problems: a value of t is needed, which depends on the sample size, which can't be solved for without specifying a value

for t, and a value for s must be used, but we don't have a value for s unless we have taken a sample.

(b) A prior estimate of σ would have to be used; otherwise the expected half-width could not be determined. With use of a prior estimate, the expected half-width is $t_{\alpha/2, 15}\, \sigma^*/4$, with σ^* denoting the prior estimate.

5.59. Show that $(n-1)s^2/\sigma^2$ has a chi-square statistic.

Solution:

This follows from Section 3.4.5.1 as we may write $W = (\frac{X-\mu}{\sigma})^2$ and recognize that W has a chi-square distribution if X has a normal distribution. Then $\sum_{i=1}^{n}(\frac{X_i-\mu}{\sigma})^2$ has a chi-square distribution with n degrees of freedom, and $\sum_{i=1}^{n}(\frac{X_i-\overline{X}}{\sigma})^2$ has a chi-square distribution with $(n-1)$ degrees of freedom. We may write this as $\dfrac{\sum_{i=1}^{n}(X_i-\overline{X})^2}{\sigma^2}$, with the numerator equal to $(n-1)s^2$.

5.61. The heat involved in calories per gram of a cement mixture is approximately normally distributed. The mean is supposed to be 90 and the standard deviation is known to be (approximately) 2. If a two-sided test were performed, using $n = 100$, what value of $\overline{x}$ would result in a p-value for the test of .242? What decision should be reached if this p-value resulted from the test?

Solution:

A p-value of .242 for a two-sided test implies that the Z-value must be 1.17 (since the cumulative probability must be $1 - .242/2 = .879$.) Then it is matter of solving $1.17 = \dfrac{\overline{X}-90}{\frac{2}{\sqrt{100}}} \Rightarrow \overline{X} = 90.234$. The decision would be to not reject the null hypothesis.

5.63. A machine is supposed to produce a particular part whose diameter is in the interval (0.0491 inch, 0.0509 inch). To achieve this interval, it is desired that the standard deviation be at most 0.003 inch. It is obvious that some machines cannot meet this requirement and will be replaced. It is not clear whether one particular machine should be replaced or not as a sample of 100 observations had a standard deviation of 0.0035 inch. This is higher than the target value but the difference could be due to sampling variation. Test the null hypothesis that the standard deviation for this machine is at most 0.003, bearing in mind that management has decided that a p-value less than .01 will motivate them to replace the machine, and assuming that evidence suggests that the part diameters are approximately normally distributed.

Solution:

The test statistic is $\chi^2 = \frac{(n-1)S^2}{\sigma_0^2} = \frac{99(.0035)^2}{(.0030)^2} = 134.75$. This is a one-sided test and $P(\chi_{n-1}^2 > 134.75 \mid \sigma^2 = \sigma_0^2) = .0098$, so there is evidence that the standard deviation has increased and that the machine should be replaced.

5.65. Assume that a 95% confidence interval for μ with σ assumed known is of width W. What would be the width of a 99% confidence interval for μ, using the same data, as a function of W?

Solution:
The multiplier of W is $2.576/1.96 = 1.314$.

5.67. An experimenter performs a hypothesis test for the mean of an assumed normal distribution, using a one-sided test (greater than). If σ is assumed to be 4.6, the mean of a sample of 100 observations is 22.8, and the p-value for the test is .06681, what was the sign of $(\bar{x} - \mu)$ and what was the magnitude of the difference?

Solution:
The p-value of .06681 corresponds to a cumulative probability of $1 - .06681 = .93319 \Rightarrow Z = 1.500$. Since the p-value is less than .5, the sign of $(\bar{x} - \mu)$ must have been positive, and using $Z = \frac{\bar{x} - \mu}{\frac{\sigma}{\sqrt{n}}}$, we can solve for the magnitude of the difference. That is, $1.5 = \frac{\bar{x} - \mu}{\frac{\sigma}{\sqrt{n}}} = \frac{\bar{x} - \mu}{\frac{4.6}{\sqrt{100}}}$, so the magnitude of the difference is 0.69.

5.69. Assume that a confidence interval for the mean is to be constructed and the t-value should be used instead of the Z-value because σ is unknown. Consider the difference between the t-value and the Z-value for a 95% confidence interval versus the difference between the two for a 90% interval. Which difference will be greater? For which will the percentage difference, relative to the t-value, be the larger? What does this suggest, if anything, about using Z-values as approximations for t-values when confidence intervals are constructed?

Solution:
The difference will be greater for the 95% confidence interval and the percentage difference will also be greater for the 95% interval. It is best to use the t-distribution whenever possible, especially for 90% intervals.

5.71. The specifications for a certain type of surveillance system state that the system will function for more than 12,000 hours with probability of at least .90.

Fifty such systems were checked and eight were found to have failed before 12,000 hours.   These sample results cast some doubt on the claim for the system.

(a) State H_0 and H_a and perform the appropriate test after first stating what the assumptions are for the test, if any.

(b) Construct the confidence bound that corresponds to this test and comment on the results.

Solution:

(a) $H_0: p \geq .90$ and $H_a: p < .90$. $Z = \dfrac{\hat{p} - p_0}{\sqrt{\dfrac{p_0(1 - p_0)}{n}}} = \dfrac{.625 - .90}{\sqrt{\dfrac{.90(1 - .90)}{8}}} = -2.59$.

The associated p-value is quite small ($.0048$), so we reject H_0.

(b) The 99% upper bound on p is found by using

$$\frac{\hat{p} + \frac{Z_\alpha^2}{2n} + Z_\alpha \sqrt{\frac{\hat{p}(1 - \hat{p})}{n} + \frac{Z_\alpha^2}{4n^2}}}{1 + Z_\alpha^2/n}$$

from Eqs. (5.4).   Substituting .625 for $\hat{p}$, 2.326 for $Z_{\alpha/2}$ and $n = 8$, we obtain .886 as the upper bound.   The confidence bound also leads us to reject the claim.

5.73. A 95% confidence interval is constructed for μ as a means of testing a null hypothesis.   If the hypothesized value was exactly in the center of the interval, what would have been the value of the test statistic if a hypothesis test had been performed?

Solution:
The value of the test statistic would have to be zero because the only way for the hypothesized value to be in the center of the interval would be for $\bar{x}$ to equal μ, which would make the numerator of the test statistic zero.

5.75. Several years ago D. J. Gochnour and L. E. Mess of Micron Technology, Inc. patented a method for reducing warpage during application and curing of encapsulated materials on a printed circuit board.   Assume that the company has found that the warpage with the standard method has had a mean of 0.065 inch and before the method was patented the scientists wanted to show that their method has a smaller mean. They decide to construct a one-sided confidence bound, hoping that the bound will be less than 0.065. They took a sample of 50 observations and found that $\bar{x} = 0.061$, with $s = 0.0018$.

(a) What assumption(s), if any, must you make before you can construct the bound? Since assumptions must be tested, can the assumption(s) be tested with 50 observations? Explain.

(b) Make the necessary assumption(s), construct the bound, and draw a conclusion.

Solution:
(a) Normality must be assumed; this can be tested with 50 observations although the test will not be very sensitive with that number of observations.

(b) Because of the sample size, we might use Z instead of t and compute the upper bound on the mean as $\bar{x} + Z_\alpha \frac{\sigma}{\sqrt{n}} = 0.061 + 1.645 \frac{0.0018}{\sqrt{50}} = 0.0614.$
Since this is less than 0.065, the new method seems superior.

5.77. Indoor swimming pools are known to have poor acoustic properties. The goal is to design a pool in such a way that the mean time that it takes a low-frequency sound to die down is at most 1.3 seconds with a standard deviation of at most 0.6 second. Data from a study to test this hypothesis were given by Hughes and Johnson's article, "Acoustic Design in Nanatoriums" (*The Sound Engineering Magazine*, pp. 34-36, April 1983). The simulated data were as follows:

1.8	2.8	4.6	5.3	6.6
3.7	5.6	0.3	4.3	7.9
5.0	5.3	6.1	0.5	5.9
2.5	3.9	3.6	2.7	1.3
2.1	2.7	3.8	4.4	2.3
3.3	5.9	4.6	7.1	3.3

Analyze the data relative to the stated objective.

Solution:
A test for normality looks okay, but that doesn't matter much because the objectives are not even close to being met, as the mean is 3.97 and the standard deviation is 1.89. The p-value for the test of the mean is .000 and the test for the variance of .36 also expectedly has a p-value of .000.

5.79. Consider Example 5.4 in Section 5.3.7. Would you have proceeded differently in working that problem? In particular, would you have made the distributional assumption that was made in that example or would you have tested that assumption? Would you have proceeded differently in terms of the distributional assumption if the objective had been to construct a confidence interval or bound on σ^2 rather than a confidence bound on the mean? Explain.

Solution:
In general, a sample of size 100 will be more than enough to permit an assumption that the distribution of the sample mean is approximately normal and to proceed to

construct a confidence interval for the population mean. Such a sample size may not be adequate when the distribution of the individual observations is highly skewed, however, and since the random variable is lifetime, it would have been a good idea to look at a histogram and/or normal probability plot of the data. If a confidence interval for σ^2 were being constructed, we would want the distribution of the random variable to be very close to a normal distribution, which wouldn't likely be the case in this application.

5.81. Explain why a p-value is *not* the probability that the null hypothesis is correct.

Solution:
A p-value cannot be the probability that the null hypothesis is correct because the null hypothesis is a statement that is either correct or not correct (and we know that null hypotheses are almost always incorrect), whereas a p-value is a non-trivial probability of some event occurring. The "probability", loosely speaking, of the null hypothesis being correct is either zero or one, and is almost certainly zero.

5.83. In a random sample of 400 industrial accidents for a large company over a period of time, it was found that 231 accidents were due to unsafe working conditions. Construct a 99% confidence interval for the true proportion of accidents due to unsafe conditions for that time period. Address the validity of this interval if these 400 accidents occurred over a period of, say, 25 years.

Solution:

$$U.L. = \frac{\hat{p} + \frac{z_{\alpha/2}^2}{2n} + z_{\alpha/2}\sqrt{\frac{\hat{p}(1-\hat{p})}{n} + \frac{z_{\alpha/2}^2}{4n^2}}}{1 + z_{\alpha/2}^2/n}$$

$$= \frac{.5775 + \frac{(2.575)^2}{2(400)} + 2.575\sqrt{\frac{(.5775)(1-.5775)}{400} + \frac{(2.575)^2}{4(400)^2}}}{1 + (2.575)^2/400}$$

$$= .639$$

$$L.L. = \frac{\hat{p} + \frac{z_{\alpha/2}^2}{2n} - z_{\alpha/2}\sqrt{\frac{\hat{p}(1-\hat{p})}{n} + \frac{z_{\alpha/2}^2}{4n^2}}}{1 + z_{\alpha/2}^2/n}$$

$$= \frac{.5775 + \frac{(2.575)^2}{2(400)} - 2.575\sqrt{\frac{(.5775)(1-.5775)}{400} + \frac{(2.575)^2}{4(400)^2}}}{1 + (2.575)^2/400}$$

$$= .513$$

This interval wouldn't have any meaning if the data were collected over a long time period, such as 25 years, because p will not be constant for a long time period. We must assume that we are constructing a confidence interval for a stable parameter.

5.85. Assume that a large-sample test of H_0: $\mu = 35$ against H_a: $\mu < 35$ is performed and $Z = 1.35$. What is the p-value?

Solution:
The p-value is $P(Z < 1.35) = .9115$.

6

Confidence Intervals and Hypothesis Tests -- Two Samples

6.1. Show that t and t^*, as discussed in Section 6.1.2, are equal when the sample sizes are equal. Does this mean that the tests are equivalent when the sample sizes are equal? Explain.

Solution:

The denominator of t^* is $\sqrt{\dfrac{s_1^2}{n_1} + \dfrac{s_2^2}{n_2}}$, with this becoming $\sqrt{\dfrac{s_1^2 + s_2^2}{n}}$ when the sample sizes are equal. The denominator of the independent-sample t-test with equal sample sizes can be written as $\sqrt{\dfrac{s_1^2 + s_2^2}{2}\left(\dfrac{1}{n} + \dfrac{1}{n}\right)}$, which is obviously equal to the denominator of t^*. The tests are not equivalent because the degrees of freedom will differ unless $s_1^2 = s_2^2$, which is very unlikely.

6.3. Western Mine Engineering, Inc. provides considerable data on mining operations at its website. For example, for 2001 there was sample data on 194 U.S. mines, 77 of which were union mines. Of the 58 union mines at which there was a wage increase, the average increase was 2.4%. Of the 70 non-union mines at which there was a wage increase, the average increase was 2.7%. Would it be possible to test whether the wage increase was equal for 2001 for the two types of mines using one of the methods given in this chapter? Why or why not? If possible, perform the test and draw a conclusion.

Solution:
It is not possible to make that test using one of the methods given in the chapter, based on the information that is given. Percentage increases are almost certainly not going to be even approximately normally distributed, and the sample sizes are not large enough, especially for the union mines, to be able to proceed without distributional information. Of course, information on variability would also have to be provided.

6.5. A two-sample t-test is used to test H_0: $\mu_1 = \mu_2$ versus H_a: $\mu_1 < \mu_2$, with $\alpha = .05$. If $n_1 = 14$, $s_1^2 = 6$, $n_2 = 15$, $s_2^2 = 7$, and it is assumed that $\sigma_1^2 = \sigma_2^2 = \sigma^2$, what is the numerical value of the estimate of σ^2?

Solution:

$$\hat{\sigma}^2 = s_{pooled}^2 = \frac{(n_1-1)s_1^2 + (n_2-1)s_2^2}{n_1 + n_2 - 2} = \frac{13(6) + 14(7)}{13 + 14} = 6.52$$

6.7. There are various ways to construct a normal probability plot, depending on what is plotted. One approach is to plot the normal score (i.e., Z-score) on the vertical axis and the observations on the horizontal axis. Another approach, illustrated in Section 6.1.2, is to have the cumulative probability on the vertical axis. Explain why a line fit through the points on a plot using either approach could not have a negative slope.

Solution:
The plotted points must slope upward because the cumulative probability by definition must increase as the variable increases, and similarly, z-scores increase as the variable increases.

6.9. A sample is obtained from each of two (approximately) normal populations and the following results are obtained: $n_1 = 8$, $\bar{x}_1 = 60$, $s_1^2 = 20$, $n_2 = 12$, $\bar{x}_2 = 75$, $s_2^2 = 15$.

(a) What would be the numerical value of the estimate of $\mu_1 - \mu_2$?

(b) Is it necessary to assume that the populations each have a normal distribution in order to obtain the estimate? Why or why not?

Solution:
(a) The estimate of $(\mu_1 - \mu_2)$ is $\bar{x}_1 - \bar{x}_2 = 60 - 75 = -15$.

(b) No, this estimate is not based on any distributional assumption.

6.11. Assume that two "treatments" (such as two blood pressure medications) are to be compared in a study with each of 20 people receiving each of the two medications. Explain why the appropriate *t*-test to use for this analysis is <u>not</u> the one that has $n_1 + n_2 - 2 = 38$ degrees of freedom.

Solution:
The samples are dependent because the same people are being measured in each of the two samples.

6.13. What is the numerical value for the degrees of freedom of an independent sample *t*-test if $n_1 = 35$ and $n_2 = 30$?

Solution:
$35 + 30 - 2 = 63$

6.15. Gunst (*Quality Progress*, October 2002, Vol. 35, pp. 107-108) gave "16 mutually independent resistivity measurements for wires from two supposedly identical processes" and showed that the *p*-value for testing that the two process means were equal was .010. The data are given below.

Process 1	Process 2
0.141	0.145
0.138	0.144
0.144	0.147
0.142	0.150
0.139	0.142
0.146	0.143
0.143	0.148
0.142	0.150

(a) Considering the small number of observations from each population, what is the first step you would take before you performed the hypothesis test?

(b) Take the action that you indicated in part (a) and then, if possible, perform the pooled-t test and state your conclusion. If it is not possible to perform that test, is there another test that could be used? If so, perform the test and state your conclusion.

Solution:
(a) Approximate normality for each set of process measurements should be checked.

(b) Normal probability plots suggest approximate normality for each process. The assumption of equality of variances for the pooled-t test must also be checked and neither Levene's test nor the F-test leads to rejection of the hypothesis of equal variances. Therefore, it is appropriate to use a pooled-t test and doing so produces a t-value of -2.98 and a p-value of .01. Thus, there is evidence that the means for the two processes are different.

6.17. Using the data in Example 6.7 in Section 6.4, construct a 99% two-sided confidence interval for σ_1^2 / σ_2^2 using the standard, F-statistic approach. Would it be possible to use the Brown-Forsythe approach with the information that is given? Why or why not?

Solution:

$$P(F_{\alpha/2, n_2-1, n_1-1} \frac{s_1^2}{s_2^2} \le \frac{\sigma_1^2}{\sigma_2^2} \le \frac{s_1^2}{s_2^2} F_{1-\alpha/2, n_2-1, n_1-1}) = 1 - \alpha, \text{ so}$$

$$F_{.005, 99, 99} \frac{16.2}{15.0} \le \frac{\sigma_1^2}{\sigma_2^2} \le \frac{16.2}{15.0} F_{.995, 99, 99} \text{ and thus } U.L. = \frac{16.2}{15.0}(1.68536) =$$

1.82 and $L.L. = \frac{16.2}{15.0}(0.59334) = 0.64.$

The Brown-Forsythe modification could not be used because absolute deviations from the means were not given.

6.19. The U. S. Consumer Product Safety Commission stated in a 11/5/01 memorandum (http://www.cpsc.gov/library/toydth00.pdf) that of the 191,000 estimated toy-related injuries in 2000, males were involved in 68% of them. Would it be possible to construct a 95% confidence interval for the difference in the proportion of males and females who were involved in toy-related injuries that year? If so, construct the interval. If not, explain why the interval cannot be constructed.

Solution:
No, such a confidence interval could not be constructed for various reasons, one of which is the fact that the data are population data, not sample data, for the year 2000.

6.21. Assume that an experimenter wishes to test the equality of two means but fears that the populations may be highly nonnormal. Accordingly, he takes samples of $n = 25$ from each population and constructs a normal probability plot for each sample. The points on one of the plots practically form a straight line, but there is a strong S-shape on the other plot.

(a) What does an S-shape suggest regarding the tails of the distribution?

(b) Despite the signal from one of the plots, the experimenter decides to proceed with the test and will reject the null hypothesis if the absolute value of the t-statistic is greater than 2. Does the plot configuration make this more likely or less likely to occur than if normality existed for each population? Explain.

Solution:
(a) This suggests that the tails are heavier than the tails of a normal distribution.

(b) Since extreme values are more likely with heavy-tailed distributions, a significant result is more likely to occur, although it is possible, but unlikely, that extreme values in opposite directions could be offsetting.

6.23. (Harder problem) Explain, in specific terms, why an inequality in an alternative hypothesis cannot be reversed before the value of the test statistic is computed if the data suggest that the direction of the inequality is incorrect.

Solution:
Doing so would bias the results. The probabilities for Type I and Type II errors hold before a sample is obtained. If it is necessary to reverse the direction of the inequality, this shows that either (a) the experimenter does not understand the subject matter very well if the absolute value of the test statistic is large, or (b) if the test statistic differs very little from zero, the "wrong sign" can occur just because of sampling variation, with no evidence that the null hypothesis should be rejected.

6.25. Oxide layers on semiconductor wafers are etched in a mixture of gases to achieve the desired thickness. Low variability of the thickness of the layers is highly desirable, with variability of the thickness being a critical characteristic of the layers. Two different mixtures of gases are under consideration and there is interest in determining if one is superior to the other in terms of variability. Samples of 25 layers are etched with each of the two mixtures, with a standard deviation of 1.85 angstroms observed for the first mixture and 2.02 angstroms for the second mixture. Test the appropriate hypothesis under the assumption that approximate normality is a plausible assumption for each population.

Solution:

We will use an F-test because of the approximate normality. Thus, $F = \dfrac{1.85^2}{2.02^2} =$ 0.84. The p-value, which is $P(F < 0.84)$, is .1649. Therefore, there is no evidence that the first mixture of gasses is superior to the second one.

6.27. If the lower limit of a (two-sided) 95% confidence interval for $\mu_1 - \mu_2$ using t is 23.4, will the limit of a lower one-sided 95% confidence interval for $\mu_1 - \mu_2$ (using the same data) be less than 23.4 or greater than 23.4? Explain.

Solution:
The lower one-sided limit will be greater than 23.4. This is because the t-value is smaller since .05 is in the lower tail of the distribution for the one-sided interval, whereas .025 is the lower-tail area for the two-sided interval. Thus, the t-value that is used for the one-sided interval is not as extreme as the one used for the two-sided interval, so the subtraction from $\bar{x}_1 - \bar{x}_2$ that is of the form $ts\sqrt{\dfrac{1}{n_1} + \dfrac{1}{n_2}}$ is smaller for the one-sided interval.

6.29. Explain why the numerator of the test statistic of the paired-t test is equivalent to the numerator of the independent sample t-test, but the denominators are not the same.

Solution:
The numerators are the same because $\bar{d}$ in the paired test equals $\bar{x}_1 - \bar{x}_2$, which is the numerator in the independent sample test. The denominators differ because with the independent test $Var(\bar{x}_1 - \bar{x}_2) = Var(\bar{x}_1) + Var(\bar{x}_2)$, whereas with the paired test, $Var(\bar{d}) = Var(\bar{x}_1 - \bar{x}_2)$, but this does not equal the sum of the two variances since the data are paired (i.e., not independent).

6.31. Explain why we do not have to assume that $\sigma_1^2 = \sigma_2^2$ whenever we use t^* (as given in Section 6.1.2) in constructing a confidence interval for $\mu_1 - \mu_2$.

Solution:

It isn't necessary to assume equal variances when t^* is used. In fact, it is used when the variances are believed to be unequal.

6.33. The Dean of Student Affairs at a large midwestern university decided to conduct a study to determine if there was a difference in academic performance between freshmen male students who do and do not have cars on campus. A sample of 100 male students from each of the two populations is selected with the null hypothesis being that the mean GPA is the same for both populations, and the alternative hypothesis is that car owners have a lower GPA. The sample without cars had a GPA of 2.7 with a variance of 0.36, as compared with the car owners, whose average GPA was 2.54 with a variance of 0.40. The alternative hypothesis is the one that the dean suggested. Perform the appropriate test and reply to the dean. Given your result, if you have constructed a confidence interval or confidence bound on the mean GPA for car owners minus the mean GPA for students who do not have a car on campus, what can you say about how the interval would appear, relative to zero? (Don't construct the interval or bound.)

Solution:
Although the distributions of GPAs for the two classes were not stated, the sample sizes are large and we would not expect either distribution to be highly skewed. Therefore, invoking the Central Limit Theorem seems reasonable, and using Z as a substitute for t seems reasonable since the t-statistic would have 198 degrees of freedom and its critical value would thus be very close to the critical value of Z. Thus,

$$Z = \frac{2.70 - 2.54}{\sqrt{\frac{0.40}{100} + \frac{0.38}{100}}} = 1.83$$

which has a p-value of .0332 for this one-sided test. Thus, there is moderate evidence to suggest that students without cars on campus do have higher GPAs. Let μ_1 denote the GPA of all students who do *not* operate a car on campus and let μ_2 denote the average GPA of all students who do operate a car on campus. We would want to construct a lower bound on $\mu_1 - \mu_2$ and if it's a 95% bound, it will be greater than zero (anything up to a 96.67% lower bound will be greater than zero).

6.35. Consider Example 6.5 in Section 6.3 and construct a 99% confidence interval for the difference of the two proportions. We know, however, what the relationship will be between the two endpoints of the interval and zero. What will be that relationship?

Solution:
Since the sample proportions differ greatly, the confidence interval should be constructed using

$$\hat{p}_1 - \hat{p}_2 \pm Z_{\alpha/2} \sqrt{\frac{\hat{p}_1}{n_1} + \frac{\hat{p}_2}{n_2}}$$

$$.35 - .23 \pm 2.576 \sqrt{\frac{(.35)(.65)}{100} + \frac{(.23)(.77)}{100}}$$

so the interval is $(-.04, .28)$. (Since this is the interval for the difference of two proportions rather than the interval for a single proportion, a negative lower limit does make sense.)

6.37. The management of a new company is trying to decide which brand of tires to use for its company cars when the tires currently on the cars have to be replaced. One of the executives stumbles onto a research report that provided the results of an accelerated testing study. (Accelerated testing is discussed in Section 14.3.) The study involved 100 tires of each of two types and showed that Brand A had an average useful life of 32,400 miles with a variance of 1,001,034 miles, whereas Brand B had an average useful life of 29,988 miles with a variance of 888,698 miles. Construct a 99% confidence interval for the mean of Brand A minus the mean of Brand B and draw a conclusion regarding which brand the company should select.

Solution:
Since each sample is large, we will use

$$(\bar{x}_1 - \bar{x}_2) \pm Z_{\alpha/2} \sqrt{s_1^2/n_1 + s_2^2/n_2}$$
$$= (32,400 - 29,988) \pm 2.575 \sqrt{1,001,034/100 + 888,698/100}$$
$$= 2,412 \pm 2.575 (137.468)$$

so that $U.L. = 2,765.98$ and $L.L. = 2,058.02$. The interval does not include zero, nor even come close to doing so. Therefore, the company should select Brand A.

6.39. (MINITAB required) One of the sample datasets that comes with MINITAB is VOICE.MTW, which is in the STUDENT1 subdirectory. A professor in a theater arts department was interested in seeing if a particular voice training class improved a performer's voice quality. Ten students took the class and six judges rated each student's voice on a scale of 1-6, before and after the class. One method of analysis would be to obtain an average score for each student, before and after the class. What would be lost, if anything in using such an approach? Perform the appropriate test with these data, after first stating the assumptions for the test and whether or not those assumptions can be tested for this dataset. What do you conclude?

Solution:

One potential problem with averaging the scores is that there could be some extreme scores, which would suggest throwing out the highest and lowest score, which of course is what is frequently done, especially in sports. Looking at the judges' scores, there are two scores that are clearly out of line, as judge #5 gave student #3 a before score of 2, whereas the other five judges gave scores of 5, 5, 6, 4, and 5, respectively. Similarly, on student #8, judge #1 gave a before score of 2, whereas the other five scores were 4, 5, 5, 4, and 5, respectively. The first judge also gave the first student an after score of 3, whereas the other scores were 5, 5, 6, 5, and 5, respectively.

Oddly, as the computer output for the paired-*t* test below indicates, the average score before the voice training (averaging over the judges and students) was greater than the average after the training. Since 6 is supposed to be the best score, this raises questions regarding the data (hypothetical or real?) and the scoring (did the judges understand that 6 is the best score rather than 1?). Presumably we can rule out that this is "bad voice training". So we have a significant result with a *p*-value of .045 but the cause of this result would have to be investigated

```
Paired T-Test and CI: Before, After

Paired T for Before - After

               N      Mean      StDev     SE Mean
Before         10     3.85000   1.04217   0.32956
After          10     3.40000   0.94428   0.29861
Difference     10     0.450000  0.610100  0.192931

95% CI for mean difference: (0.013561, 0.886439)

T-Test of mean difference = 0 (vs not = 0):
T-Value = 2.33  P-Value = 0.045
```

6.41. A new employee at a company performs a paired-*t* test on some company data with 15 pairs and finds that all of the differences have a positive sign, with an average of 4.9 and a standard deviation of 1.8.

(a) What decision should be reached regarding the test if the alternative hypothesis is $H_a: \mu_d < 0$?

(b) A person who is familiar with the data is surprised at the conclusion drawn by the new employee and contends that the differences should have been negative. What is the likely cause of this discrepancy?

Solution:
(a) The null hypothesis would not be rejected.

(b) The wrong difference was probably formed (e.g., $X_2 - X_1$ should have been formed instead of the reverse).

6.43. A paired-t two-sided test is performed and the p-value is .04, with the value of the test statistic being positive. What would be the relationship between the endpoints of a 95% confidence interval for μ_d and zero if the interval had been constructed using the same data?

Solution:
Since the test statistic was positive and the p-value was less than $1 - .95 = .05$, it follows that both endpoints of the interval would exceed zero.

7

Tolerance Intervals and Prediction Intervals

7.1. Which of the intervals presented in this chapter, if any, are of the general form $\hat{\theta} \pm t\, s_{\hat{\theta}}$? What is an interval called that is of this form?

Solution:
None of the intervals in the chapter are of that form because that is the form of the confidence interval for a parameter and confidence intervals were not covered in the chapter.

7.3. When will a prediction interval and a tolerance interval be the same and when will they differ?

Solution:
As the previous problem illustrates, they are the same in the parameters-known case, but will be different when parameters must be estimated, as then the tolerance interval has a degree of confidence and a percent coverage, which automatically separates it from a prediction interval.

7.5. Will a prediction interval for a single observation always be wider than a confidence interval for μ (using the same data)? Explain.

Solution:
Yes, the prediction interval will always be wider because there is an additional element of uncertainty since the interval is for a random variable rather than a fixed parameter. More explicitly, $s\sqrt{1 + 1/n}$ is larger than $s/\sqrt{n}$ for any value of n.

7.7. Which of the following intervals would not be appropriate for compliance monitoring: (a) confidence interval for μ, (b) prediction interval, (c) tolerance interval?

Solution:
(a) and (b). Interest should center on estimating the percentage of observations that fall outside the accepted range of values. Neither a confidence interval for the mean nor a prediction interval would be of any help in this regard.

7.9. A steel mill has been asked to produce high-strength low-alloy steel (H.S.L.A.) that has a minimum yield strength of 60,000 psi. As the steel is produced, each coil is tested for strength. The product is a thin sheet steel that is several feet wide and several thousand feet long. The strip has been rolled into coils to make them easier to handle. The production department is anxious to send out a large shipment to a particular customer. There isn't time to test every item but you are given the task of determining if the minimum yield strength requirement has been met for each item. Would you use a method given in this chapter for addressing this issue? If so, how would you proceed? If not, explain what you would do.

Solution:
One possibility would be to construct a confidence interval on the proportion of observations that fall below the minimum yield strength requirement.

7.11. Explain to someone unfamiliar with tolerance intervals the difference between a 99% tolerance interval that has a confidence coefficient of 95% and a 95% tolerance interval that has a confidence coefficient of 99%.

Solution:
A 99% tolerance interval is an interval designed to cover 99% of the population values. This cannot be done with certainty, however, nor would we expect the interval to cover exactly 99% of the population, so a degree of confidence is attached to the interval, with the degree of confidence indicating the probability that the interval contains, in this case, *at least* 99% of the population values. A 95% tolerance interval with 99% confidence is one for which the user is 99% confident that the constructed interval contains at least 95% of the population.

7.13. A toothpaste was formulated with a silica content of 3.5% by weight. A manufacturer randomly samples 40 tubes of toothpaste and finds that the average silica content is 3.6% with a standard deviation of 0.28%. You are an employee of this company and you are asked to construct a 95% prediction interval, using this information (only). Could you do so using methodology presented in this chapter? If so, compute the interval; if not, explain why not.

Solution:
No, not using methodology presented in the chapter. Since a prediction interval is sensitive to non-normality, regardless of the sample size, non-normality must be considered, and here we have data involving percentages.

7.15. Very large amounts of data are now being routinely maintained in datafiles. Assume that a company has such a file for a manufactured part, with the distribution being approximately normal with a mean of 0.82 inch and a standard deviation of 0.06 inch.

(a) What would be the endpoints of an approximate 99% tolerance interval?

(b) Why would this interval not have a degree of confidence attached to it, considering the information given in the problem?

Solution:
  (a) The interval is given by $\mu \pm Z_{\alpha/2}\,\sigma = 0.82 \pm 2.58\,(0.06) = (0.67, 0.97)$.

(b) There is no degree of confidence attached to the interval in part (a) because μ and σ are assumed known.

7.17. Assume that a company constructed a 95% prediction interval for the number of future failures of a particular component by a specified time. Assume that the distribution of failure times is close to being symmetric and there is subsequently an increase in the mean number of failures after the interval has been constructed, due to a process control problem. What action should be taken regarding the interval that was constructed?

Solution:
A new interval should be constructed using recent data.

7.19. Assume that a 99% tolerance interval to contain at least 95% of a population with a normal distribution has been constructed and shortly thereafter quality improvement efforts were successful in reducing the standard deviation. How will this affect the percentage of the population covered by the interval and the degree of confidence? Or will only one of these be affected, or will neither be affected? Explain.

Solution:
Since a normality based tolerance interval is of the form $\bar{x} \pm ks$, if σ is reduced, the proportion of the population that is covered by the interval will increase. If we knew the percent reduction in σ, then we could determine the percent increase in the coverage.   Of course nearly equal k values can be found for different combinations of degree of confidence and percent coverage, but it would seem to be most appropriate to view the degree of confidence being fixed and the percent coverage having increased.

7.21. Is it possible to solve explicitly for the sample size in a prediction interval, given the desired width of the interval, the degree of confidence, and an estimate of sigma? Explain.

Solution:
No, this cannot be done explicitly because a t-variate must be used, which depends upon the sample size. The sample size could be solved for explicitly only if an assumed value of σ were used.

8

Simple Linear Regression, Correlation and Calibration

8.1. Consider the four datasets given in the datafile ANSCOMBE.MTW, which is also one of the sample datasets that comes with MINITAB. Construct a scatter plot for each dataset and then obtain the regression equation for each dataset, using MINITAB or other software. Comment. In particular, would we really want to use simple linear regression for each of these datasets? Which datasets, if any, would seem to be suitably fit by a simple linear regression equation?

Solution:
A straight-line relationship is suggested only for the first and third datasets, with the relationship being weak for the first dataset and an obvious outlier present in the third dataset.

8.3. Assume that $\hat{\beta}_1 = 2.38$ in a simple linear regression equation. What does this number mean, *in words*, relative to X and Y?

Solution:
The number 2.38 is the estimate of the change in Y per unit change in X, for the range of X-values in the sample data used to obtain the regression equation.

8.5. Show that the expected value of a residual is zero when the simple linear regression model is the correct model. (*Hint*: Write the ith residual as $Y_i - \hat{Y}_i$ and then obtain the expected value.)

Solution:
By definition, $E(Y_i) = \beta_0 + \beta_1 X_i$ since $E(\epsilon_i) = 0$. $E(\hat{Y}_i) = E(\hat{\beta}_0 + \hat{\beta}_1 X_i) = E(\bar{Y} + \hat{\beta}_1(X_i - \bar{X})) = E(\bar{Y}) + X_i E(\hat{\beta}_1) - \bar{X} E(\hat{\beta}_1)$, assuming X to be fixed.

$E(\bar{Y})$ is the average of the $E(Y_i)$, which is $\beta_0 + \beta_1 \bar{X}$. Since $E(\hat{\beta}_1) = \beta_1$ under the assumption that the model is correct (as is to be shown in Exercise 8.38), it follows that $E(\hat{Y}_i) = \beta_0 + \beta_1 X_i = E(Y_i)$. Thus, $E(Y_i - \hat{Y}_i) = E(Y_i) - E(\hat{Y}_i) = 0$.

8.7. Show that $s_{\hat{\beta}_1} = s/\sqrt{S_{xx}}$. [*Hint*: First show that $\hat{\beta}_1$ can be written as $\sum k_i Y_i$ with $k_i = (X_i - \overline{X})/S_{xx}$.]

Solution:

By definition, $\hat{\beta}_1 = S_{xy}/S_{xx}$, with $S_{xy} = \sum xy - \dfrac{(\sum x)(\sum y)}{n}$. The latter can be written as $\sum(x - \overline{x})(y - \overline{y}) = \sum(x - \overline{x})y$ since $\overline{y}\sum(x - \overline{x}) = 0$. Thus, $\hat{\beta}_1$ can be written as $\hat{\beta}_1 = \dfrac{\sum(x-\overline{x})y}{S_{xx}} = \sum k_i y_i$ with $k_i = \dfrac{x-\overline{x}}{S_{xx}}$. Accordingly, $Var(\hat{\beta}_1) =$

$Var(\sum k_i y_i) = \sum k_i^2 Var(y_i) = \sigma^2 \sum k_i^2$ since $Var(y_i) = \sigma^2$ for all i. $\sum k_i^2 = \dfrac{S_{xx}}{(S_{xx})^2}$

$= \dfrac{1}{S_{xx}}$, so $Var(\hat{\beta}_1) = \dfrac{\sigma^2}{S_{xx}}$, as was to be shown.

8.9. Critique the following statement: "The least squares estimates minimize the sum of the squares of the errors."

Solution:
The least squares estimates minimize the sum of the squares of the *residuals*. The errors would be known only if the parameters were known, in which case they would of course not be estimated.

8.11. The data in EX8-12ENGSTAT.MTW consist of two columns of 100 random numbers, each generated from the standard normal distribution. Thus, there is no relationship between the numbers, but when either column of numbers is used to represent the predictor variable and the other is used to represent the dependent variable, the p-value for testing H_0: $\beta_1 = 0$ is .026 (i.e., significant at the .05 level), but the value of R^2 is only .049 (i.e., 4.9%).

(a) What does this suggest about looking at p-values in regression, especially when n is large?

(b) Use the second column of numbers in the file to represent the dependent variable and construct a scatter plot. Does the plot suggest any relationship between the two "variables"?

(c) Can the numerical value of the t-statistic for testing H_0: $\beta_1 = 0$ be determined from what has been given in the problem statement (i.e., not using the data)? If so, what is the value? If not, explain why the value can't be computed.

(d) Can the absolute value of this t-statistic be determined? If so, what is the value?

Solution:
(a) Paraphrasing what some prominent statisticians have stated, "You can prove anything with enough data". Here we see an example of that as the p-value is

small, but the R^2 value is also small. The p-value in this case simply gives the likelihood of $\beta_1 = 0$, which of course isn't likely, but rejecting this hypothesis, which is likely to happen if the sample is large enough, doesn't rule out the possibility of β_1 being very close to zero.

(b) There is no discernible relationship, although curiously, there is evidence of nonconstant variance even though these are just random numbers.

(c) No, the value cannot be determined from what is stated in the problem, but it can be gleaned from the scatter plot in additional to what is stated in the problem.

(d) Yes, the combination of the p-value and the sample size does specify the absolute value of the t-statistic, which is 2.26.

8.13. What does $1 - R^2$ mean in words?

Solution:
This is the percent of the variation in Y that is <u>not</u> explained by the regression model.

8.15. Given the following numbers, fill in the blank:

Y	3	5	6	7	2	4	8
$\widehat{Y}$	2.53	___	5.30	6.68	3.22	5.30	8.06

Could the prediction equation be determined from what is given? Explain why or why not. If possible, give the equation.

Solution:
The number must be 3.88 in order for the residuals to sum to zero. There is not enough information given to determine the prediction equation

8.17. Consider the following data.

X	1 2 3 4 5 6 7
Y	5 6 7 8 7 6 5

First obtain the prediction equation and compute R^2. *Then*, graph the data. Looking at your results, what does this suggest about which step should come first?

Solution:
The prediction equation is $\widehat{Y} = 6.29$ because $\widehat{\beta}_1 = 0$. The graph shows a quadratic relationship between X and Y, with no evidence of linearity being part of that

relationship. This illustrates that two-dimensional data must first be graphed before a regression model is fit.

8.19. For the data in the file 75-25PERCENTILES2002.MTW, regress the 75th percentile SAT scores against acceptance rate and then plot the standardized residuals against $\hat{Y}$. Compare your plot with Figure 8.4. Can you see from your graph what can be seen from Figure 8.4?

Solution:
The plot of the standardized residuals against $\hat{Y}$ is given below. It is the mirror image of Figure 8.4 because the coefficient of acceptance rate is negative. Of course the nonconstant variance is evident here, as it was in Fig. 8.3; we simply have to reverse our view.

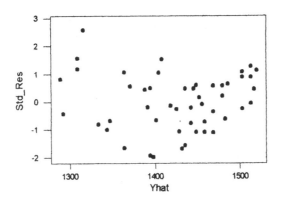

8.21. It is possible to see the effect of individual observations dynamically by using one of the available Java applets. As of this writing there are several such applets available on the Internet. One of these is given at the following URL: http://gsbwww.uchicago.edu/fac/robert.mcculloch/research/teachingApplets/Leverage/index.html . Another one is available at http://www.stat.sc.edu/~west/javahtml/Regression.html. Use one of these two applets to show what happens when the rightmost point is moved up or down. Then show what happens when a point near the center, in terms of the x-coordinate, is moved up and down. Explain why the effects differ.

Solution:
(student computer exercise)

8.23. Consider a scatter plot of Y against X in simple linear regression. If all of the points practically form a line with a negative slope, what would be the approximate value of the correlation coefficient (i.e., r_{xy})?

Solution:
The correlation coefficient would be approximately − 1.

8.25. Explain what the word "least" refers to in the "method of least squares".

Solution:
The word "least" refers to the fact that a minimization is taking place, and what is being minimized is the sum of the squared residuals.

8.27. Many, if not most, colleges and universities use regression analysis to predict what a student's four-year GPA would be and use this predicted value as one factor in reaching a decision as to whether or not to accept an applicant. Georgia Tech has been one such university, and in the mid-1980s they used a Predicted Grade Index (PGI) as an aid in reaching decisions. The regression equation, using their notation, was PGI $= -0.75 + 0.4$ GPA $+ 0.002(SAT\text{-}M) + 0.001(SAT\text{-}V)$. (*Source*: Student newspaper; of course, "GPA" in the equation represents high school GPA.) Bearing in mind that this was before the recentering of the SAT scores and also before grade inflation, answer the following:

  (a) What is the largest possible value of PGI?

  (b) Considering your answer to part (a) and any knowledge that you might have of the school, for what approximate range of actual college GPA values would you expect the fit to be the best?

  (c) Similarly, for what range of values of the predictors (roughly) would you expect the equation to be useful, and could we determine the set of possible values of the predictors for which the equations *could* be used?

Solution:

  (a) 3.25

  (b) Probably about 2.2-3.0.

  (c) GPA: 3.3-4.0   SAT-M: 620-730   SAT-V: 530-620

These of course are just estimates. We would need the data in order to determine the space of values for which the prediction equation could be used.

8.29. Consider the following data for a simple linear regression problem. The sample size was $n = 20$ and when H_0: $\beta_1 = 0$ was tested against H_a: $\beta_1 \neq 0$, it was found that the value of the test statistic was positive and was equal to the critical (tabular) value.

  (a) What is the numerical value of the correlation coefficient, assuming that X is a random variable?

(b) What does your answer to part (a) suggest about relying on the outcome of the hypothesis test given in this exercise for determining whether or not the regression equation has predictive value?

Solution:

(a) We first determine the value of R^2 from the equation $R^2 = t^2/(n - 2 + t^2)$, which we can do since the value of the test statistic, t, is equal to the critical (tabular) value. With $\alpha = .05$ and $n = 20$, $t_{.025, 18} = 2.1009$. Therefore, $R^2 = 2.1009^2/(20 - 2 + 2.1009^2) = .1969$. Since the value of the test statistic is positive, $\hat{\beta}_1$ is positive and hence r_{xy} is positive. So $r_{xy} = +\sqrt{R^2} = \sqrt{.1969} = .4438$.

(b) Since R^2 can obviously be low when $H_0: \beta_1 = 0$ is rejected, just rejecting the null hypothesis is not a sufficient condition for the model to have predictive value.

8.31. Consider a simple linear regression prediction equation written in an alternative way. If $\hat{\beta}_1 X$ is replaced by $\hat{\beta}_1(X - \overline{X})$, what will be the numerical value of the intercept if $\sum X = 23$, $\sum Y = 43$, and $n = 20$? Can the numerical value of $\hat{\beta}_1$ be determined from what is given here, combined with the value of $\hat{\beta}_0$? Why, or why not? If possible, what is the value?

Solution:

Since $\hat{\beta}_0 = \overline{Y} - \hat{\beta}_1 \overline{X}$, the model written in the alternative form results in $\hat{\beta}_0 = \overline{Y} = 43/20 = 2.15$. No, the value of $\hat{\beta}_1$ cannot be determined from the value of $\hat{\beta}_0$ and the other information that is given.

8.33. Hunter and Lamboy (1981, *Technometrics*, Vol. 23, pp. 323-328; discussion: pp. 329-350) presented the following calibration data that were originally provided by G. E. P. Box:

Measured Amt.	Molybdenum Known Amt.	Measured Amt.	Known Amt.
1.8	1	6.8	6
1.6	1	6.9	6
3.1	2	8.2	7
2.6	2	7.3	7
3.6	3	8.8	8
3.4	3	8.5	8
4.9	4	9.5	9
4.2	4	9.5	9
6.0	5	10.6	10
5.9	5	10.6	10

(a) Does a sufficiently strong regression relationship exist between the known and measured amounts of molybdenum for calibration to be useful? Explain.

(b) If so, construct a 95% calibration confidence interval for the known amount when the measured amount is 5.8, using Expression (8.18).

(c) Regarding Expression (8.18), what would the width of the interval approach if $\widehat{\sigma}^2$ were virtually zero? Does the limit seem reasonable?

(d) Assume that two analysts work with this set of data. If the second analyst decides to use inverse regression instead of the classical theory of calibration, will the equation for $\widehat{X}^*$ (i.e., using inverse regression) be expected to differ very much from the expression for $\widehat{X}$ (classical theory of calibration)? Explain.

Solution:
(a) Part of the regression output is as follows.

```
Regression Analysis: Measured versus Known

The regression equation is
Measured = 0.760 + 0.987 Known

Predictor      Coef    SE Coef      T       P
Constant     0.7600    0.1294     5.87   0.000
Known        0.98727   0.02085   47.35   0.000

S = 0.267838  R-Sq = 99.2%  R-Sq(adj) = 99.2%
```

Since the R^2 value is quite high, there is obviously a strong regression relationship, so calibration should be useful.

(b) The calibration interval is of the form

$$\overline{x} + \widehat{\gamma}(Y_o - \overline{Y}) \pm t_{1-\alpha/2, n-2} \sqrt{\frac{\widehat{\sigma}^2 \widehat{\gamma}^2 (1 + \frac{1}{n} + (\widehat{x}_o - \overline{x})^2 + \widehat{\sigma}^2 \widehat{\gamma}^2)}{S_{xx}(1 + \frac{\widehat{\sigma}^2 \widehat{\gamma}^2}{S_{xx}})}}$$

For a 95% calibration confidence interval, $t_{1-\alpha/2, n-2} = t_{.975, 18} = 2.10$. With Y_o = 5.8, n = 20, $\overline{x}$ = 5.5, $\overline{Y}$ = 6.19, S_{xx} = 165, $\widehat{\sigma}^2$ = 0.07, $\widehat{\gamma} = \widehat{\beta}_1 S_{xx} / \widehat{\beta}_1^2 S_{xx} + \widehat{\sigma}^2$)
= [0.987(165)]/[(0.987)^2(165) + 0.07] = 1.013, and $\widehat{x}_o = (Y_o - 0.76)/ 0.987 = $
5.11, we obtain 5.5 + 1.013 (5.8 − 6.19) ± 2.10a, with

$$a = \sqrt{\frac{0.07(1.013)^2(1 + \frac{1}{20} + (5.11-5.5)^2 + 0.07(1.013)^2)}{165(1 + \frac{(0.07)(1.013)^2}{165})}}$$

We obtain $a = 0.90$ so the interval is $5.10 \pm 2.10\,(0.90) = (3.21, 6.99)$

(c) The width of the interval would approach zero, which is as it should be, and the interval would approach the point $\bar{x} + (Y_o - \bar{Y})/\hat{\beta}_1$, which is the point estimate $\hat{x}_o$. Notice that this expression is what results from solving the simple linear regression prediction equation $\hat{Y}_o = \bar{Y} + \hat{\beta}_1(x_o - \bar{x})$, after removing the hat on $\hat{Y}_o$ and placing it on x_o since the latter is being predicted.

(d) The expressions should be very similar because of the very high correlation (.996) between the known and measured amounts.

8.35. The following calibration data were originally given by Kromer et al. (*Radiocarbon*, 1986, Vol. 28(2B), pp. 954-960).

```
Y| 8199 8271 8212 8198  8141 8166 8249 8263 8161 8163 8158
   8152 8157 8081 8000  8150 8166 8083 8019 7913
X| 7207 7194 7178 7173  7166 7133 7129 7107 7098 7088 7087
   7085 7077 7074 7072 7064  7062 7060 7058 7035
```

Each Y-value is the radiocarbon age of an artifact, with the corresponding X-value representing the age of the artifact determined by methods that are closer to being accurate. The question to be addressed is whether calibration can be used to determine an estimate of what the X-value would be if it had been obtained.

(a) What is the first step in making this determination?

(b) If practical, determine the length of a 99% calibration confidence interval when Y is 8000. If this would not be practical, explain why.

Solution:
(a) The first step is to determine if there is a strong relationship between X and Y. Whether we regress Y on X or X on Y, the R^2 value is only .443 (of course the numbers must be the same). Furthermore, even though this is data from the literature, a scatter plot does not suggest any relationship between X and Y.

(b) In view of the absence of a clear relationship between X and Y, it would be unwise to construct a calibration confidence interval as we would construct such intervals only if a relationship had been established.

8.37. Croarkin and Varner (*National Institute of Standards and Technology Technical Note*, 1982) illustrated the use of a linear calibration function to calibrate optical imaging systems. With $Y =$ linewidth and $X =$ NIST-certified linewidth, they obtained a regression equation of $\hat{Y} = 0.282 + 0.977X$. With $n = 40$, $S_{xx} = 325.1$, and $s = 0.0683$, determine each of the following.

(a) What is the numerical value of the t-statistic for testing the hypothesis that $\beta_1 = 0$?

(b) Is the t-statistic large enough to suggest that the equation should be useful for calibration?

(c)   What is the expression for $\widehat{X}_0$, assuming that the classical theory of calibration is used?

Solution:

(a) The t-statistic is $t = \dfrac{\widehat{\beta}_1}{\dfrac{s}{\sqrt{S_{xx}}}} = \dfrac{0.977}{\dfrac{0.0683}{\sqrt{325.1}}} = 257.92.$ This value is far in excess of any value for t that might be considered the smallest value in order for the regression equation to be useful.

(b) Yes, this is a very large value of the t-statistic, so the equation should be useful for calibration.

(c) Solving for X from $\widehat{Y} = 0.282 + 0.977X$, we obtain $\widehat{X} = (Y - 0.282)/0.977 = -0.289 + 1.024\,Y$.

8.39. Construct an (x, y) scatter plot that shows very little linear relationship between X and Y when X varies from 10 to 20 but does show evidence of a linear relationship when the range is increased to 10-30. Would you fit a (common) simple linear regression line through the entire set of data or would you fit a line to each segment? (The latter is called *piecewise linear regression*.) Which approach would provide the best fit?

Solution:
Preferably, a line should be fit to the data for X from 20 to 30. If the entire range is to be used, different lines should be fit from 10 to 20 and from 20 to 30. It is important to recognize, however, that lines or planes should not be fit in parts of the data space where there is no clear relationship.

8.41. In production flow-shop problems, performance is often evaluated by minimum makespan, this being the total elapsed time from starting the first job on the first machine until the last job is completed on the last machine. We might expect that minimum makespan would be linearly related, at least approximately, to the number of jobs. Consider the following data, with X denoting the number of jobs and Y denoting the minimum makespan in hours.

X	3	4	5	6	7	8	9	10	11	12	13
Y	6.50	7.25	8.00	8.50	9.50	10.25	11.50	12.25	13.00	13.75	14.50

(a) From the standpoint of engineering economics, what would a nonlinear relationship signify?

(b) What does a scatter plot of the data suggest about the relationship?

(c) If appropriate, fit a simple linear regression model to the data and estimate the increase in the minimum makespan for each additional job. If doing this would be inappropriate, explain why.

Solution:
(a) Minimum makespan scheduling is an objective that can be met using local search algorithms. A nonlinear relationship with Y increasing at an increasing rate for certain values of X might suggest that improvement should be sought by using better algorithms, or the increase may simply be a characteristic of the system.

(b) The scatter plot suggests a linear relationship.

(c) $R^2 = .996$, so the linear relationship is obviously quite strong. $\hat{\beta}_1 = 0.82273$, so this is the estimate of the additional minimum make-span for each additional job, within the range of the number of jobs (X) in the dataset.

8.43. Assume a set of 75 data points such that $\hat{Y} = -3.5 + 6.2X + 1.2X^2$. Would it make sense to report the value of r_{xy}? Why or why not? If the answer is yes, can the value of r_{xy} be determined from the regression equation? Explain.

Solution:
The value of r_{xy} could certainly be reported, but it would be more meaningful to give the value of R^2. The value of r_{xy} cannot be determined from the prediction equation.

8.45. What is a consequence, if any, of trying to fit a simple linear regression model to a data set that has a considerable amount of pure error?

Solution:
The consequence is simply that there won't be a very good fit since there is no way to accommodate pure error.

8.47. A student takes the data in 75-25PERCENTILES2002.MTW that were used in Section 8.2.5 and also in Chapter 1. The student decides to use the 25th percentile score as the dependent variable, and the 75th percentile score as the independent variable, which is the reverse of what was done in Section 8.2.5. The student fits the model and computes R^2. Will it be the same value as was obtained in Section 8.2.5? Why, or why not?

Solution:
The value of R^2 will be the same because R^2 in the one-predictor case is simply the square of the correlation coefficient, and $r^2_{XY} = r^2_{YX}$.

8.49. A simple linear regression equation is obtained, with $\hat{Y} = 2.8 - 5.3X$ and $R^2 = .76$. What is the numerical value of r_{xy}?

Solution:

$r_{xy} = -\sqrt{R^2} = -0.87$, the sign being negative because $\hat{\beta}_1$ is negative.

8.51. If you were asked to suggest a transformation of the predictor based on Figure 8.6, what would it be (if any)? Does Figure 8.7 suggest the same transformation that you are recommending (if any)? Explain. Do you believe that either of these two figures is misleading? Explain.

Solution:
Fig. 8.7 suggests that a quadratic term should be used in the model, whereas Fig. 8.6 suggests the use of a reciprocal term.

8.53. Consider the regression equation $\hat{Y} = 13.4 + 6.8\,X + 2.3\,X^2$. Can we infer from the magnitude of the regression coefficients that the linear term is approximately three times as important in predicting Y as the quadratic term? Explain.

Solution:
No, we can't make such a statement. First, we cannot, in general compare the magnitudes of regression coefficients for regressors that are not orthogonal. Another problem is that the units are different If the squared term has very large numbers, the coefficient will often be quite small. Then if we simply compared the magnitude of the regression coefficients, we might conclude, for example, that the linear term is 10,000 times more important than the quadratic term, which would be nonsensical if there was say, a 2% increase in R^2 due to the quadratic term being added to the model.

8.55. Use any regression applet that is available on the Internet, such as one of the two given in Exercise 8.21, to see what happens when you have several points that plot as a very steep line with a positive slope and then you add a point that is far to the right and below these points. See how the line changes as you move that point down and away from the line.

Solution:
The reader who does the exercise will observe that the line will follow the movement of the point.

8.57. The correlation between SAT-Math and SAT-Verbal scores, using the average scores for each of the 50 states as one data point, is quite high (over .90). Is either one of these variables realistically the "dependent" variable and the other the "independent" variable in a regression model? If not, how can the high correlation be explained?

Solution:

Neither variable would be a natural choice as the dependent variable. The high correlation can be explained by the fact that both variables are strongly related to overall intelligence.

8.59. Proceed as follows to demonstrate the difference between statistical significance in simple linear regression and practical significance, analogous to the discussion in Section 5.9. For one data set, use the integers 20-30 for one set of X-values. Then generate 11 random errors from the standard normal distribution and add those to the X-values to create the Y-values. Then repeat using the integers 10-40 so that 31 random errors must be generated. Perform the regression analysis on each data set and compare the two values of the standard error of $\widehat{\beta}_1$. Compare the two values of s relative to the difference in the two standard errors of $\widehat{\beta}_1$ and relative to the two t-statistics for testing that $\beta_1 = 0$. What have you demonstrated?

Solution:
The results for one simulation are shown below.

X- values	s	standard error of $\widehat{\beta}_1$	t
20-30	0.89	0.085	12.48
10-40	1.19	0.024	42.51

The same relationship between Y and X exists for the two sets; the only difference is the spread of the X-values. The larger spread results in a larger value of s, but also a larger t-statistic (and a larger R^2 value). The larger spread has resulted (in this particular simulation) in a worse fit as measured by s (and of course what is being estimated is $\sigma = 1.00$) but a much better fit according to the t-statistic. This shows the effect of the increased spread and how "greater statistical significance" can be influenced by the spread.

8.61. Consider the following statement: "I know that $Var(\widehat{\beta}_1)$ is affected by the spread of the predictor values, so I am going to use a measurement unit of inches instead of feet for the model predictor so as to increase the spread." Will this work? Show what will happen algebraically when feet are used initially and then inches are used as the measurement unit. Comment.

Solution:
As one might expect, this won't help. Although $\sqrt{Var(\widehat{\beta}_1)}$ is reduced by 12 in the conversion to inches, $\widehat{\beta}_1$ is reduced by the same amount. Thus, the parameter when the unit of measurement is inches is estimated with the same precision as the parameter when the unit of measurement is feet.

8.63. To see the connection between the least squares regression equation and the equation for a straight line from algebra, consider two points (x_1, y_1) and (x_2, y_2). Show that the least squares solution for $\widehat{\beta}_1$ is mathematically equivalent to

$(y_2 - y_1)/(x_2 - x_1)$, which of course in algebra is the slope of the line. What is the numerical value of R^2 for this line?

Solution:
Using Eq. (8.8), we obtain the numerator of the slope as $\dfrac{X_1Y_1 + X_2Y_2 - X_1Y_2 - X_2Y_1}{2}$, which can be written as $\dfrac{(Y_2 - Y_1)(X_2 - X_1)}{2}$, after simple algebraic simplification. The denominator of the slope can be similarly simplified to $\dfrac{(X_2 - X_1)^2}{2}$. Thus, the numerator divided by the denominator is $\dfrac{(Y_2 - Y_1)}{(X_2 - X_1)}$. The value of R^2 is 1.0 because the line passes through both points and is thus an exact fit to the data.

8.65. J. A. Hoeting and A. R. Olsen gave a table in their article, "Are the Fish Safe to Eat? Assessing Mercury Levels in Fish in Maine Lakes" (in *Statistical Case Studies: A Collaboration Between Academe and Industry*, R. Peck, L. D. Haugh, and A. Goodman, eds., ASA-SIAM series) in which the p-value for a simple linear regression model is .0002 but the R^2 value is only .13. The messages are thus conflicting as to whether or not the model is an adequate fit to the data. Explain how this could happen.

Solution:
These conflicting messages will often occur with datasets that are at least moderately large and that analysis had $n = 119$. The standard error of $\hat{\beta}_1$ is greatly affected by the sample size, so here the small p-value is misleading because of the moderately large sample size. The value of R^2 will also often be small when n is large.

8.67. The dataset SUGAR.MTW that comes with MINITAB was analyzed in the context of a paired-t test in Section 6.2. The objective required more than the use of a paired-t test, however, since there was a desire to see if there is a relationship between the expensive lab measurements and the less expensive field measurements so that what the lab measurement would be if made can be estimated from the field measurement. The results for the regression of field measurement on lab measurement are shown below.

```
Regression Analysis: Field versus Lab

The regression equation is
Field = 0.261 + 0.0536 Lab

Predictor     Coef    SE Coef     T       P
Constant   0.261046  0.002802  93.17   0.000
Lab        0.053554  0.005222  10.25   0.000
```

```
S = 0.00901890   R-Sq = 80.2% R-Sq(adj) = 79.4%
```

This suggests a moderately strong linear relationship between the two sets of measurements, although the strength of the relationship might not be sufficient for the intended use of the equation. Now construct a scatter plot of field versus lab. What lesson have you learned?

Solution:
The scatterplot given below shows that a linear regression model with only a linear term should not be fit as clearly there is no straight-line relationship.

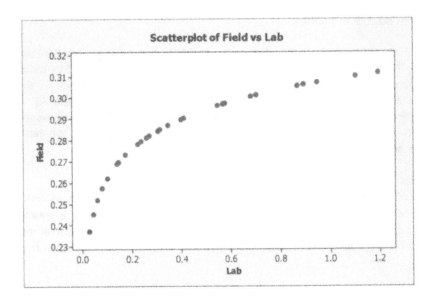

9

Multiple Regression

9.1. It was stated in Section 9.3 that for the example in that section a normal probability plot of the residuals and a plot of the standardized residuals against the predicted values did not reveal any problems with the assumptions, nor did those plots spotlight any unusual observations. Construct those two plots. Do you agree with that assessment?

Solution:
The standardized residuals plot does not exhibit anything unusual, as can be seen below.

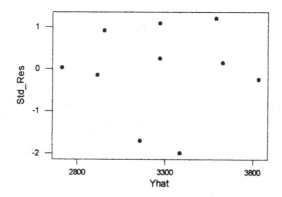

The normal probability plot, as follows, does not provide strong evidence of nonnormality.

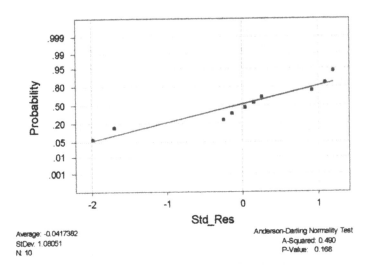

Normal Probability Plot

Average: -0.0417382
StDev: 1.08051
N: 10

Anderson-Darling Normality Test
A-Squared: 0.490
P-Value: 0.168

9.3. Notice that the plot of score against reputation in Figure 9.1 exhibits curvature. Does this suggest that a nonlinear term in reputation should be included in the model? Fit the model (the data are in RANKINGS2002.MTW) with linear terms in all predictors shown in Figure 9.1 and construct a partial residuals plot for reputation and compare the message that plot gives compared with the message of the scatter plot. Comment.

Solution:
No, although the plot of score against reputation exhibits some curvature in the lower left corner, the plot should not be relied on since it ignores the existence of the other predictors, which do not all have small pairwise correlations. The partial residual plot given below does not exhibit curvature. Furthermore, when a quadratic term in reputation is created and used as a candidate variable in a stepwise regression, the quadratic term is not selected for the model.

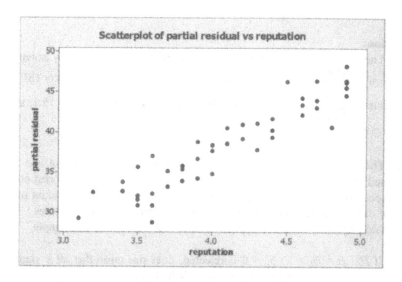

Scatterplot of partial residual vs reputation

9.5. The following output was obtained using the "trees" data that comes with the MINITAB software, with the objective being to predict the volume of a tree from knowledge of its diameter and height (in feet), with the latter being easier to determine than the volume.

```
Regression Analysis: Volume versus Diameter, Height

The regression equation is
Volume = - 58.0 + 4.71 Diameter + 0.339 Height

Predictor    Coef     SE Coef       T         P
Constant   -57.988     8.638     -6.71     0.000
Diameter     4.7082    0.2643    17.82     0.000
Height       0.3393    _____     2.61     0.346

S = 3.882    R-Sq = _____     R-Sq(adj) = 94.4%

              Analysis of Variance
Source            DF     SS       MS       F       P
Regression         2   7684.2   3842.1   _____   0.000
Residual Error    28    421.9   _____
Total             30   8106.1
```

(a) Fill in the four blanks.

(b) When regression is used for prediction, it is generally for predicting what will happen in the future. If, however, in this application we know the diameter and height of a particular tree, are we trying to predict the future height of that tree, or another tree, or neither tree? Explain.

Solution:
(a) The standard error for the coefficient of height is found by solving the equation $\frac{0.33893}{SE} = 2.61$, which leads to 0.1302 for the standard error (SE). The R^2 value is $\frac{7684.2}{8106.1} = 94.8\%$, the mean square value is $\frac{421.9}{28} = 15.1$, and the F-statistic is $\frac{3842.1}{15.1} = 254.97$.

(b) The diameter of a tree can be easily obtained and the height can also be obtained. The volume of a tree is another matter. The Allegheny Forest obviously has more than 31 trees so the objective would be to estimate the amount of timber in the forest, or to estimate the amount of timber in specific trees. So the regression equation would be applied to trees that were not in this sample.

9.7. If $H_0: \beta_1 = \beta_2 = \dots \beta_k = 0$ is rejected, does this mean that all k regressors should be used in the regression equation? Why or why not?

Solution:
If the hypothesis is rejected, it means that at least one of the available regressors should be used in the model. The next step is to determine which one(s) should be used.

9.9. Assume that two predictors are used in a regression calculation and $R^2 = .95$. Since R^2 is so high, could we simply assume that both predictors should be used in the model? Why or why not?

Solution:
No, one of the two could be of little value. Since there are only two predictors and the R^2 value is high, the t-statistics can be relied on as an indicator of whether both variables or only one should be used in the equation, assuming that the statistics are interpreted properly.

9.11. Assume that a multiple linear regression equation is determined using two independent variables. Sketch an example of how we would expect a plot of the standardized residuals against X_i to look if the model is a reasonable proxy for the true model and the regression assumptions are met.

Solution:
The plotted points would be such that that the configuration of points could be enclosed in a rectangle.

9.13. A regression equation contains two regressors, both of which are positively correlated with Y. Explain how the coefficient of one of the regressors could be negative.

Solution:
The sign of each of the two regression coefficients is determined by the relationship between the correlation of Y with each of the regressors and the correlation between the regressors. A negative coefficient could easily result if one of the regressors was a much weaker predictor than the other one and the regressors were at least moderately correlated.

9.15. Explain why a transformation of one or more of the predictors is preferable to transforming Y when trying to improve the fit of a regression model.

Solution:
If the assumptions are approximately met, this will be lost if Y is transformed. This is why model improvement should usually center on predictor transformations.

9.17. D. A. Haith (1976, "Land Use and Water Quality in New York Rivers," *Journal of the Environmental Engineering Division,* American Society of Civil Engineers, **102** (EE1), Paper 11902, February 1-15, 1976) investigated the relationship between water quality and land usage. He sought to determine if relationships existed between land use and water quality, and if so, to determine which land uses most significantly impact water quality. If such relationships do exist, a model would be sought that would be useful "for both prediction and water quality management in northeastern river basins with mixtures of land uses comparable to the basins used in the study."

Haith stated:

> In the present case, a suitable regression would be one which explains a
> significant portion of water quality variation with a small number of land-use
> variables. In addition, these land use variables would be uncorrelated
> (independent). Such a regression would indicate the extent to which water
> quality is related to land uses and the relative importance of the various land
> uses in determining water quality.

In working this problem, you may ignore the second sentence in this quote since methods for attempting to obtain an estimate of the independent effects of correlated variables are not well known and are beyond the scope of this text. You are welcome to address the issue, however, if desired.

Haith used correlation and regression analyses and cited papers that described similar analyses of water quality data. He gave nitrogen content as one measure of water quality, with one set of land-use variables given as: (1) percentage of land area currently in agricultural use; (2) percentage of land use in forest, forest brushland, and plantations; (3) percentage of land in residential use; and (4) percentage of land area in commercial and manufacturing use. Haith also broke land-use variables into various subcategories, but we will be concerned only with the data set that has the four land-use variables.

The land-use information for each river basin was obtained from New York's Land Use and Natural Resource Laboratory. Additional background information can be found in Haith (1976). You are encouraged, but not required, to read the latter. Discussions of the Haith paper were given by Chiang in the October issue of the same journal and by Shapiro and Küchner in the December issue.

The data are as follows.

River Basin	Total Nitrogen[a]	Agriculture Area[b]	Forest (%)	Residential (%)	Commercial-Industrial (%)
Olean	1.10	530	26	63	1.2	0.29
Cassadaga	1.01	390	29	57	0.7	0.09
Oatka	1.90	500	54	26	1.8	0.58
Neversink	1.00	810	2	84	1.9	1.98
Hackensack	1.99	120	3	27	29.4	3.11
Wappinger	1.42	460	19	61	3.4	0.56
Fishkill	2.04	490	16	60	5.6	1.11
Honeoye	1.65	670	40	43	1.3	0.24
Susquehanna	1.01	2,590	28	62	1.1	0.15
Chenango	1.21	1,830	26	60	0.9	0.23
Tioughnioga	1.33	1,990	26	53	0.9	0.18
West Canada	0.75	1,440	15	75	0.7	0.16
East Canada	0.73	750	6	84	0.5	0.12
Saranac	0.80	1,600	3	81	0.8	0.35
Ausable	0.76	1,330	2	89	0.7	0.35
Black	0.87	2,410	6	82	0.5	0.15
Schoharie	0.80	2,380	22	70	0.9	0.22
Raquette	0.87	3,280	4	75	0.4	0.18
Oswegatchie	0.66	3,560	21	56	0.5	0.13
Cohocton	1.25	1,350	40	49	1.1	0.13

[a] Total nitrogen is given as mean concentration in milligrams per liter.
[b] Area is given in square kilometers.

What model would you recommend to accomplish Haith's objective, and would you recommend using all of the data points?

Solution:
As a side note, this dataset is analyzed extensively on pages 468-477 of *Modern Regression Methods* by T. P. Ryan. That analysis indicated the relative importance of the land uses by employing some moderately sophisticated techniques. The ordering was Residential, Agricultural, Commercial-Industrial, and Forest. From inspection of the data, we see that the Hackensack river basin stands out from the others because it is easily the smallest in area and its residential usage percentage is an order of magnitude larger than all of the other percentages. The Neversink river basin is also a potentially influential data point and we might question it

simply because of the unusual name. The best subset model is probably (Residential, Forest) which has an R^2 value of .86 when Hackensack is excluded from the calculations. Some improvement is realized when log(Forest) is used instead of Forest as R^2 increases to .88.

9.19. Consider the following data set for two predictors.

Y	5	3	8	7	5	4	3	7	5	4
X_1	5	3	5	5	4	3	2	6	4	7
X_2	2	1	3	2	4	3	2	3	2	5

(a) Construct the regression equation with only X_2 in the model and note the sign of the slope coefficient.

(b) Then put both predictors in the model and compare the sign of the coefficient of X_2 with the sign when only X_2 is in the model. Explain why the sign changes.

(c) Would you recommend that the model contain either or both of these predictors, or should a nonlinear form of either or both of them be used?

Solution:

(a) The regression equation is $\widehat{Y} = 4.59 + 0.190 \, X_2$.

(b) The regression equation with both predictors is $\widehat{Y} = 2.69 + 0.809 X_1 - 0.425 X_2$. The sign changes because X_1 is a much stronger predictor than X_2, and the predictors are correlated.

(c) Neither predictor should be used in linear form as neither predictor is significant when used together nor when used individually in a simple linear regression model. There is no evidence that a nonlinear term in either would be helpful as scatter plots do not suggest a nonlinear term and $X_i log(X_i)$, $i = 1, 2$, is not a significant predictor.

9.21. Choudhury and Mitra (2000, *Quality Engineering*) considered a problem in manufacturing tuners that required attention. Specifically, the design of the printed circuit boards (PCBs) had changed and the shop floor workers felt that the quality of soldering was poor. Since rework was expensive, there was a need to study the process and determine optimum values for the process variables for the new PCB design. The resultant experiment used nine controllable factors and two uncontrollable (noise) factors. (The latter are, in general, factors that are either very difficult to control during manufacture, or else cannot be controlled at all during normal manufacturing conditions but can be controlled in an experimental study such as a pilot plant study.)

Of the two uncontrollable factors, the engineers knew that the dry solder defect was the more important of the two types of defects (shorting was the other defect).

An analysis was performed for each of the two types of defects. Since the two response variables are each count variables, possible transformation of each variable should be considered since counts do not follow a normal distribution. The study was performed by taking, for each response variable, the square root of the count under each of the two levels of the uncontrollable factor and combining those two numbers into a single performance measure: $Z(y_1, y_2) = -\log_{10}\sum_{i=1}^{2}y_i^2/n$.

Converting the counts into a single performance measure is in accordance with approaches advocated by G. Taguchi (see Section 12.17), but it should be noted that the use of such performance measures has been criticized considerably during the past fifteen years.

Rather than use that approach, however, we will use a more customary approach and sum the defects under the two conditions of the uncontrollable factor and then take the square root of the sum. Since it is very unlikely that all nine factors under study are significant, we will view this as a screening experiment. The data are in CH9ENGSTATEXER20.MTW.

(a) Use all nine of the available predictors in the model and identify those variables for which the p-value is less than .05. Which variables are selected using this approach?

(b) Then use stepwise regression to identify the significant effects (use $F_{IN} = F_{OUT} = 4.0$). What variables are selected using this method?

(c) Compute R^2 for each of the two models and check the assumptions. Is either model or both models acceptable in terms of R^2? Which model do you prefer?

Solution:
(a) The only predictor with a p-value less than .05 is X_8, although X_4 is close at .053.

(b) When stepwise regression is used, X_8, X_4, and X_6 enter the model in this order and remain in the model.

(c) The model with only X_8 has an R^2 value of .558; the R^2 value for the model with the three predictors is .792. Certainly the latter is preferable and acceptable. The normal probability plot of the standardized residuals gives no evidence of nonnormality and a plot of the standardized residuals against the predicted values does not provide any evidence of nonconstant error variance or any other problem. There is also no evidence of nonnormality for the model with only X_8, but there is some evidence of nonconstant error variance when the standardized residuals are plotted against the fitted values. That is not particularly relevant here, however, because the model with three predictors is clearly the preferred model.

9.23. Apply stepwise regression (with $F_{IN} = F_{OUT} = 4.0$) to the etch rate data that were analyzed in Section 9.3. Does the stepwise algorithm select the same model that results from selecting variables for which the p-values are less than .05 when all of the variables are in the model? Explain why the t-statistics and p-values that are displayed when the stepwise algorithm is used differ from the corresponding statistics when all of the predictors are used in the model.

Solution:

No, only CF_4 flow and Power are selected by the stepwise regression algorithm, whereas all three of the predictors have a p-value less than .05 when all three are in the model. The t-statistics and p-values differ because they are based on different estimates of the error variance.

9.25. A company that has a very large number of employees is interested in determining their bodyfat percentage as simply and subtly as possible, without having to resort to underwater weighing or the use of body calipers. The reason for doing so is to identify employees who might be expected to benefit greatly from an exercise program, and who would be encouraged to participate in such a program. The company executives read that bodyfat percentage can be well-approximated as $((\text{height})^2 \ x \ (\text{waist})^2)/(\text{bodyweight} \ x \ c)$, with $c = 970$ for women and 2,304 for men. The company would like to use a linear regression equation and would also like to have some idea of the model worth. Accordingly, the company intends to select a small number of employees to use to validate the model, using underwater weighing. If the model checks out, their waist measurements could be obtained along with their height and weight during their annual physical exam. Is it possible to transform the stated model and obtain a multiple linear regression model? If so, how would you proceed to obtain the model? If it is not possible, explain why it isn't possible.

Solution:

It is possible to transform the model to a multiple linear regression model if the error is multiplicative or if the error is additive but is so small that it can be effectively ignored. Specifically, taking the logarithm of each side of the equation

$$\text{Bodyfat \%} = ((\text{height})^2 \ x \ (\text{waist})^2)/(\text{bodyweight} \ x \ c)$$

produces

$$Log \ (\text{Bodyfat \%}) = 2 \ Log \ (\text{height}) + 2 \ Log(\text{waist}) - Log \ (\text{bodyweight}) - Log(c)$$

which is of the general form

$$Y = \beta_0 + \beta_1 \ \text{height} + \beta_2 \ \text{waist} + \beta_3 \ \text{bodyweight}$$

which with an additive error term is a multiple linear regression model.

9.27. Show that if the relationship $C_p = p + a$ holds for two values of p (where a is a constant), the smaller value of R^2_{adjusted} will occur at the larger of the two values of p.

Solution:
$C_p = \dfrac{SSE_p}{\hat{\sigma}^2_{full}} - (n - 2p)$ with $\hat{\sigma}^2_{full} = \dfrac{SSE_{full}}{n - a}$, with a denoting the number of available predictors. When all of the available predictors are in the model, $a = p$ and $SSE_p = SSE_{full}$. Then $C_p = (n - p) - (n - 2p) = p$.

9.29. Consider the following regression equation: $\hat{Y} = 13.4 + 6.0X_1 + 1.9X_2 - 2.6X_3 + 6.1X_4$. Looking at the equation coefficients, state the conditions, if any, under which we could conclude that X_1 is more than three times as important as X_2 in the prediction of Y, X_3 is negatively correlated with Y, and the predictive value of X_4 is approximately equal to the predictive value of X_1.

Solution:
Regarding the two pairwise comparisons of predictors, we can make those statements only if the four predictors are uncorrelated or at least have very small pairwise correlations, and the predictors have been standardized by subtracting the mean and dividing by the standard deviation, or divided by $\sqrt{S_{xx}}$. The statement about X_3 being negatively correlated with Y does not require standardization but does require that the predictors be (virtually) uncorrelated.

9.31. For the example given in Section 9.3, what would be the width of a 90% confidence interval for the bulk gas flow parameter?

Solution:
Let $\hat{\beta}_1$ denote the estimator of the bulk gas flow parameter. The width is then
$2(t_{.05,6}) \, S_{\hat{\beta}_1} = 2(1.9432)(0.6122) = 2.379$.

9.33. Consider a problem with two regressors.

(a) If there are four selected values for each regressor and a scatter plot of all combinations of the two sets of values plots as a rectangle, would it be practical to construct a 95% prediction interval for any combination of regressor values such that each value is within its extremes? Explain.

(b) Now assume that the regressors are random and the regressors have a correlation of .91. Would it still be practical to construct the prediction interval for any combination of regressor values within their respective extremes? Explain.

Solution:
(a) Yes, it would be practical because any pair of values within the individual extremes would lie within the predictor space.

(b) No, because there will be many pairs of predictor values that lie outside the predictor space.

9.35. Consider the data in Exercise 9.5. Could the value of R^2 when *only* Diameter is used in the model be determined from the output once the blanks are filled in? If so, what is the value? If not, explain why not.

Solution:
No, it cannot be determined; additional information is needed.

9.37. Consider Exercise 9.19, for which the sign of X_2 changes, depending on whether X_1 is in or out of the regression equation. Because of this, would it be better to construct a confidence interval for β_2 when only X_2 is used in the equation, or should the confidence interval be constructed at all, based on what is given in the problem?

Solution:
It would be inadvisable to construct a confidence interval for β_2 when both regressors are in the model because the coefficients are then essentially uninterpretable since the correlation between the two regressors is .586, suggesting that the regressors would have been random if these were real data. A confidence interval with X_2 the only regressor in the model would be interpretable but would not be of any practical value if other terms belonged in the model.

9.39. Explain why there are no extreme predictor values in the example in Section 9.3. Does the fact that there are no extreme predictor values preclude the possibility of having data points that could be overly influential? Explain.

Solution:
There are no extreme values in the example in Section 9.3 because an experimental design was used. Data points could still be influential, however, if there are any extreme response values.

9.41. (Harder problem) Assume that there are two predictors in a regression equation and the unit of measurement for one of them is feet. If the unit is changed to inches, what would you expect to be the effect on the regression coefficient, and would you anticipate any change in the coefficient for the other predictor?

Solution:
The change from feet to inches will cause the regression coefficient for that term to be 1/12 of its magnitude when the term was in feet. The other regression coefficient will remain unchanged. This can be explained as follows. A linear transformation of a predictor does not change the relationship of that predictor to

the response variable, nor does it change the relationship of it to the other predictor. Since these relationships remain unchanged, the only adjustment must be in the magnitude of the regression coefficient of the predictor for which there is a scale change.

9.43. Assume that you have constructed a normal probability plot of a set of standardized residuals in a regression analysis and the plot exhibits a borderline result. What action would you take?

Solution:
Since a normal probability plot of the standardized residuals is borderline, a simulation envelope should definitely be constructed. Because of the supernormality property, there is a good chance that the simulation envelope will lead to the conclusion of nonnormal errors.

9.45. Consider the following data, which are also in EX944ENGSTAT.MTW:

Row	Y	X_1	X_2
1	21.31	1.35	16.10
2	22.06	2.34	10.17
3	20.25	2.85	8.81
4	20.36	1.55	14.77
5	17.51	2.50	10.32
6	20.67	1.48	15.22
7	17.15	1.79	14.03
8	14.52	1.84	14.49
9	13.47	3.50	7.93
10	11.20	1.67	18.62
11	28.37	1.99	10.46
12	18.08	2.25	11.24
13	25.60	3.04	7.78
14	30.00	2.59	8.38
15	24.56	3.56	6.88
16	23.40	3.87	6.47
17	24.35	2.05	10.92
18	15.27	1.68	15.51
19	26.67	1.45	13.54
20	14.48	2.12	12.69

(a) Fit the model $Y = \beta_0 + \beta_1 X_1 + \beta_2 X_2$ and plot the standardized residuals against X_2; then plot the raw residuals against X_2; and finally construct the partial residual plot for X_2. Notice that the last plot differs considerably from the other two plots.

(b) What transformation is suggested by the last plot, if any? The true model contains a reciprocal term in X_2, not a linear term. Which plot gave the appropriate signal?

(c) Fit the true model and compare the R^2 value for this model with the R^2 value for the original model. Comment.

(d) What have you just demonstrated?

Solution:

(a) The standardized residuals plot, residuals plot, and partial residuals plot are as follows.

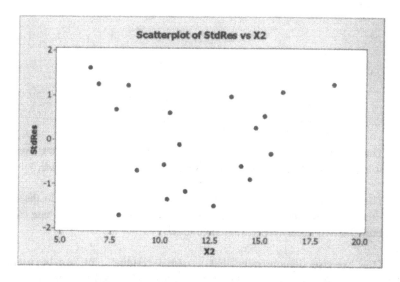

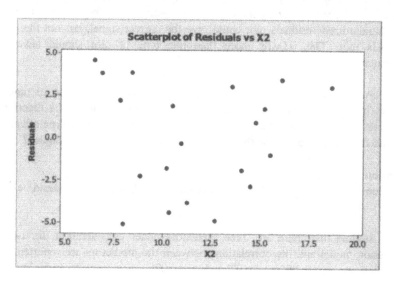

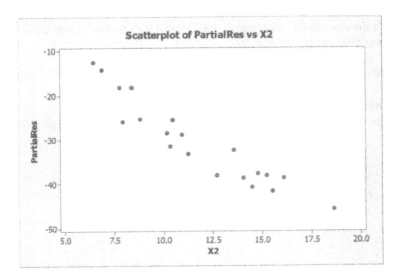

(b) The last plot suggests a reciprocal term in X_2, whereas the first two plots suggest a quadratic term. Thus, the last plot gave the appropriate signal.

(c) The R^2 value is .990 when the reciprocal term in X_2 is used in the model rather than the linear term. This is a vast improvement over the R^2 value of .618 when only the two linear terms are in the model.

(d) This exercise demonstrates that a partial residual plot can be useful in indicating how to improve a model, whereas a standardized residuals plot and a plot of the raw residuals may not do so. This won't always be the case, however, as a standardized residuals plot could give the correct signal, but not the partial residuals plot. This is why more than one type of plot should always be constructed.

9.47. An operations engineer is interested in modeling the length of time per month (in hours) that a machine will be shut down for repairs as a function of machine type (1 or 2) and the age of the machine (in years). What type of variable will machine type be and how should the values of the variable be assigned?

Solution:
Machine type is an indicator variable and the values that are used should be 0 and 1.

9.49. Assume that there are three predictors under consideration for use in a regression model and the correlations between the predictors are reported to be the following: $r_{12} = .88$, $r_{13} = .92$, and $r_{23} = .02$

(a) Would you use all three of the predictors in a regression equation, or would you need more information to make that determination? Explain.

(b) Is there any reason to believe that at least one of the reported correlations may be incorrect? Explain.

Solution:

(a) All three predictors should almost certainly not be used because of the high pairwise correlations between the first predictor and each of the other two. We don't know the extent to which these predictors are correlated with Y, however, so it is possible that none of them would be used.

(b) The first predictor is highly correlated with each of the other two predictors, but the latter are practically uncorrelated. Of course this is unlikely and it can be shown that the determinant of the correlation matrix is negative, which means that at least one of the pairwise correlations is incorrect.

9.51. (Harder problem) Assume that there are four predictors in a regression model and a partial residuals plot has been constructed for X_3. If a least squares line is fit to the points in the plot, what will be the slope and intercept of the line?

Solution:

This is a plot of $e + \hat{\beta}_3 X_3$ against X_3. Therefore, in a simple linear regression, $e + \hat{\beta}_3 X_3$ plays the role of Y and X_3 plays the role of X. The slope is S_{xy}/S_{xx} with $S_{xy} = \sum XY - \sum X \sum Y/n$ and $S_{xx} = \sum X^2 - (\sum X)^2/n$. Here this becomes $\sum XY = \sum(e + \hat{\beta}_3 X_3)(X_3) - (\sum X_3)\sum(e + \hat{\beta}_3 X_3)/n$. We have that $\sum e = 0$ from Eq. (8.7a) and $\sum eX_3 = 0$ from Eq. (8.7b). Therefore, $\sum XY = \hat{\beta}_3 \sum X_3^2 - \hat{\beta}_3(\sum X_3)^2/n = \hat{\beta}_3 S_{xx}$. Therefore, the slope is $\hat{\beta}_3$. The intercept is $\overline{Y} - \hat{\beta}_3 \overline{X} = \hat{\beta}_3 \overline{X} - \hat{\beta}_3 \overline{X} = 0$ since $\overline{e} = 0$.

10

Mechanistic Models

10.1. Assume you expect that a one-predictor model will be sufficient for predicting a particular response variable. No one with subject matter expertise is available to suggest a specific mechanistic model, but it is believed that a nonlinear relationship exists between these two variables. You have a sample of 50 (x, y) observations and are given a week to develop some type of model that gives good results. How would you proceed?

Solution:
The first step would be to construct a scatter plot of the data to confirm that there does seem to be a nonlinear relationship and that it could not be represented by a linear model with nonlinear terms. After this determination has been made, a good approach would be to try to match the scatter plot configuration with one of the scatter plots of nonlinear models given by Ratkowsky (1990, references). Nonlinear modeling software can also be used for helping narrow down the options.

10.3. (Harder problem) Taguchi (2005) stated that the volume of a tree, Y, can be estimated from the following function of chest-height diameter, D, and height, H:

$$Y = a_1 D^{a_2} H^{a_3}$$

If a_1, a_2, and a_3 are unknown, as Taguchi assumed and which seems quite likely, they would have to be estimated. Thus, a nonlinear model is being postulated, which we observe could be transformed into a linear regression model if the error is multiplicative. If the latter is not true and the error is additive, the transformation would still give good results if the error is small.

The idea of estimating the volume of trees in a forest is an important practical problem and various models have been presented in the literature for the MINITAB trees data, which is one of the sample datasets that comes with the MINITAB software. That dataset contains volume, diameter, and height data for 31 trees in the Allegheny National Forest. Volume was measured in cubic feet, the diameter was in inches (measured at 4.5 feet above the ground), and the height is in feet.

Taguchi (2005) gave the following model:

$$Y = 0.0001096 \, D^{1.839} H^{0.8398}$$

with the coefficients obtained by applying nonlinear regression to a dataset of 52 trees that were cut down. Volume was measured in cubic meters, diameter was stated to be the "chest-high diameter" and was measured in centimeters, and height was measured in meters.

Because of the pervasive influence of Taguchi, let's assume that the foresters at Allegheny National Forest decided to use an empirical-mechanistic modeling approach. That is, they would start with Taguchi's model and see how it worked with their data rather than trying to apply their meager (we shall assume for this example) knowledge of statistics.

(a) What adjustment(s), if any, would have to be made to the model given by Taguchi (2005) before it can be used to see how well it applies to the MINITAB trees data?

(b) Make any necessary adjustment(s) and compare the predicted values with the values for volume in the TREES dataset. Also compute the correlation between these two sets of values.

(c) Note the very high correlation between the two sets of values, but also note the relatively large differences between those values. To what might the high correlation and the large differences be attributed? In particular, could the difference between the "chest-high diameter" used by Taguchi and the diameter taken at 4.5 feet above the ground in the MINITAB data be a factor? Explain.

(d) What would be your recommendation to the foresters at Allegheny National Forest? In particular, would you suggest that the foresters fit a simpler multiple linear regression model obtained after taking the logarithm of volume, diameter, and height? Is this what would result from taking the logarithm of each side of the appropriate model? Explain.

Solution:
(a) Either the model could be converted to fit the data or the data could be converted so that it confirms to the units used in the model. From the standpoint of model validation, the latter would be more logical since datasets could have different combinations of units. Therefore, for model validation with the MINITAB trees dataset, Height is multiplied by 0.3048 to convert the measurement to meters; Diameter is multiplied by 2.54 to convert it to centimeters, and Volume is multiplied by $(.3048)^3$ to convert to cubic meters. These conversions lead to the comparison of predicted and observed volume measurements in part (b) for the 31 trees.

(b)

Volume	Predicted	Volume	Predicted
0.29166	0.38965	0.95711	1.03199
0.29166	0.39084	0.77588	1.10238
0.28883	0.39716	0.72774	0.99104
0.46440	0.61480	0.70509	0.92054
0.53236	0.70268	0.97693	1.11606
0.55784	0.72960	0.89764	1.17017
0.44174	0.62251	1.02790	1.13897
0.51537	0.69306	1.08454	1.33395
0.63996	0.74394	1.20630	1.46037
0.56351	0.71641	1.56875	1.70015
0.68527	0.76070	1.57725	1.75446
0.59465	0.74839	1.65087	1.79138
0.60598	0.74839	1.45832	1.80983
0.60315	0.72383	1.44416	1.80983
0.54085	0.81333	2.18040	2.48877
0.62863	0.91861		

The correlation between these two sets of values is .988.

(c) Every predicted value exceeds the corresponding observed value by a sizable amount, so this is a poor fitted model when applied to the MINITAB trees data. Certainly, the difference between chest height and 4.5 feet could be a factor, but more than likely the difference is due to different types of trees. (Of course other questions that can be raised include whether the data used to develop the model is fictitious, is there measurement error in that dataset and/or the MINITAB dataset, are there other possible problems with the MINITAB dataset, etc.)

(d) The general form of the model stated in Taguchi (2005) might be used as a starting point. If the general form is appropriate and the model error is either multiplicative or additive but small, then a linear model (which is easier to work with than a nonlinear model) with log terms in Y, D, and H should provide a good fit to the data. The analysis of this dataset is discussed extensively in Section 6.6.4 of *Modern Regression Methods* by T. P. Ryan and this is one of the models that is discussed. Other models are also discussed, relative to the model assumptions. The reader is referred to that discussion.

10.5. Wolfram (*A New Kind of Science*, Wolfram Media, 2002) makes some provocative statements regarding mechanistic models in the section "Ultimate Models for the Universe" in Chapter 9, Fundamental Physics, of his book. In particular, he states that "since at least the 1960s mechanistic models have carried the stigma of uninformed amateur science." In continuing to discuss physics, Wolfram states: "And instead I believe that what must happen relies on the phenomena discovered in this book -- and involves the emergence of complex properties without any obvious underlying mechanistic physical set up." Since the author is one of the world's leading scientists and is the developer of the very popular *Mathematica* software, such statements should not be taken lightly. Do you adopt a counterviewpoint in regard to the use or possible use of mechanistic models in your field? Explain.

Solution:
 (student exercise)

11

Control Charts and Quality Improvement

11.1. Assume that a practitioner decides to use a control chart with 2.5-sigma limits. If normality can be assumed (as well as known parameter values), what is the numerical value of the in-control ARL?

Solution:
$P(Z > 2.5)$ is .0062, with $Z \sim N(0,1)$. Therefore, the probability of a point falling outside the limits when the process is in control is $2(.00621)$, so the in-control ARL is $1/(2(.00621)) = 80.5$.

11.3. Assume that you have 200 observations on customer waiting time and you want to construct a control chart to see if there is any evidence that the process of waiting on customers was out of control for the time period covered by the data. What is the first thing that you should do?

Solution:
The first step should be to look at the distribution of the 200 numbers and, in particular, determine if the distribution appears to be close to a normal distribution.

11.5. An R-chart is constructed with subgroups of size 5 being used. Assuming normality of the individual observations, what will be the lower control limit if 3-sigma limits are used, if indeed there will be a lower limit? What would you recommend to the person who constructed this chart?

Solution:
There will not be a 3-sigma lower control limit for subgroups of size 5. The user would be much better off using probability limits, which will ensure a lower control limit.

11.7. In a study that involved the salaries of chemical engineers, the following information is available, through July 1, 2000 (*Source*: Webcam):

Number of Firms Responding: 159
Number of Employees: 1,715
Mean Average Minimum Salary: $46,732
Mean Average Salary (June 30, 2000): $61,261
Mean Average Maximum Salary: $75,843

This is how the data were reported.

(a) Would you suggest any changes? In particular, what is meant by "mean average salary"? Is that the same as "average salary"? Explain.

(b) Can you estimate σ from these data using the range approach, even though ranges are not explicitly given? Explain.

(c) Assume that you wanted to use a Stage 1 control chart approach to analyze these data so as to try to identify any companies that are unusual in terms of the variability of their salaries, as well as their average salaries. Could any of the control charts presented in this chapter be used for this purpose? Why or why not? Could this be accomplished if you had the entire dataset and you sampled from that dataset? If so, how would you perform the sampling?

Solution:

(a) Depending upon whether there is primary interest in extremes of average company salaries or extreme individual salaries, it might be preferable to report the smallest and largest salaries over all of the firms, but not if the companies are of primary interest. The "mean average salary" must be the overall average salary, averaging over the firms.

(b) We could do so only in a rough way, by focusing on the distribution of $\overline{X}$ rather than the distribution of X. The range of the averages would be a rough estimate of $6\sigma/\sqrt{n}$. We don't have a single, constant value of n, so the best we can do is divide the total number of employees by the number of firms. Doing so produces $1725/159 = 10.85$, so we will use $n = 11$. Then $6\sigma/\sqrt{11} = 75,843 - 46,732 = 29,111$. Thus, $\hat{\sigma} = \dfrac{29,111\sqrt{11}}{6} = 16,091.7$. This is a large number, but the difference of the largest and smallest average salaries is also large.

(c) An $\overline{X}$ chart and an s-chart could be used to look for firms that are extreme in their average salary and salary variability, respectively. Regarding sampling, the 159 firms that responded does constitute a sample. Since the spotlight is on extreme firms, Analysis of Means (ANOM), presented in Section 12.2.3, might also be used.

11.9. Explain why it is difficult to control variability when individual observations are obtained rather than subgroups.

Solution:

Individual observations obviously don't contain any information about variability separate from information on the process average, unlike subgroup data. Creating artificial subgroups for the purpose of computing a measure of variability is of very little value.

11.11. If quality improvement is the goal, as it should be, should all attributes control charts have a LCL? Explain.

Solution:

All attributes control charts that are used should indeed have a LCL. Without it, it can be difficult to determine when quality improvement has occurred whenever a Shewhart attributes control chart is used.

11.13. Assume that a basic CUSUM procedure is used for individual observations and the first three Z-values are 0.8, 0.4, and 1.3, respectively. Assuming that $k = 5$ is used and both sums had been set to zero, what are the three values of S_H and S_L that correspond to these Z-values? What do these Z-values and your CUSUM values suggest about the process?

Solution:
The correspondence is as follows:

Z	S_H	S_L
0.8	0.3	0
0.4	0.2	0
1.3	1.0	0

S_L remains at 0 because the Z-values are positive. The S_H values are obtained by subtracting 0.5 from each Z-value and adding the preceding S_H value.

11.15. Compute the value of the process capability index C_{pk} when the engineering tolerances are given by LSL = 20 and USL = 40, and assume that μ = 30 and σ = 10/3.

Solution:

Since μ is equidistant from LSL and USL, we may use either of these in computing C_{pk}. Using the USL, we obtain $Z = (40 - 30)/(10/3) = 3$. $C_{pk} = 1/3$. $Z_{min} = (1/3)(3) = 1$.

11.17. A department manager wishes to monitor, by month, the proportion of invoices that are filled out correctly. What control chart could be used for this purpose?

Solution:

A p-chart should be used, preferably with regression-based limits.

11.19. Assume that a control chart user fails to test for normality and independence before using an $\overline{X}$-chart. Assume further that the data are independent but have come from a nonnormal distribution. If the control limits that are constructed turn out to be at the 0.01 and 98.2 percentiles of the (nonnormal) distribution of $\overline{X}$, what is the in-control ARL?

Solution:

The in-control ARL is $1/(.0001 + (1 - .982)) = 55.2$.

11.21. The UCL on an np-chart with 3-sigma limits is 39.6202 and the midline is 25. What would be the LCL on the corresponding p-chart if $n = 500$?

Solution:

The LCL on the np-chart must be $25 - (39.6202 - 25) = 10.3798$. The LCL on the p-chart must thus be $10.3798/500 = 0.0207596$.

11.23. Determine the 3-sigma control limits for an $\overline{X}$-chart from the following information: subgroup size = 4, number of subgroups = 25, $\overline{\overline{X}} = 56.8$, and $\overline{s} = 3.42$.

Solution:

The $\overline{X}$-chart control limits would be obtained as $\overline{\overline{X}} \pm 3\dfrac{(\overline{s}/c_4)}{\sqrt{n}}$, with the value of c_4 obtained from Table F with $n = 4$. The control limits would thus be obtained as $56 \pm 3[\dfrac{(3.42/0.9213)}{\sqrt{4}}]$. So, LCL = 50.43 and UCL = 61.57.

11.25. Assume that an $\overline{X}$-chart is constructed using 3-sigma limits and parameter values that are assumed to be known, and the assumption of normality is met. If the process mean remains constant, what is the probability that the next plotted point will be outside the control limits?

Solution:

The probability is .0027 since this is the sum of the two tail areas beyond the 3-sigma limits.

11.27. Consider the baseball data in the file ERAMC.MTW that was used in Exercise 1.1.

(a) Construct the normal probability plot and note that the hypothesis of normality is rejected if we use $\alpha = .05$ since the p-value for the test that MINITAB uses is .018. Would you suggest that an X-chart be constructed for these data? Why or why not?

(b) Even though doing so might be highly questionable in view of the normal probability plot, construct an X-chart for American League earned run averages (ERAs), using the data from 1901 to 1972. (Note that "averages" are being treated as individual observations since the averages are computed over teams and there is no interest in the individual teams.) Use the moving range approach to estimate sigma.

(c) Would you have constructed the control limits using a different approach? Explain. If you would have constructed the control limits differently, do so. Are any points outside the control limits now?

(d) Now plot the earned run averages starting in 1973, one at a time in simulating real time. Construct an X-chart with 3-sigma limits, and a CUSUM chart using $k = 0.5$ and $h = 5$. Is there a sufficient number of observations to enable the parameters to be estimated for the X-chart and have the chart perform in a reliable manner? Similarly, would you expect the CUSUM procedure to be reliable (i.e., have a small performance variance)?

(e) Do the two chart methods give different signals after 1972? Since the designated hitter rule was started in 1973, which method gives the correct signal?

Solution:

(a) The normal probability plot is given below. This or a histogram shows the left skewness. It would be inadvisable to construct an X-chart with such strong evidence of nonnormality.

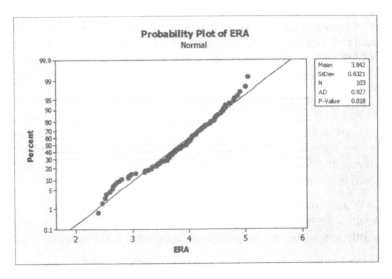

(b) The *X*-chart is given below.

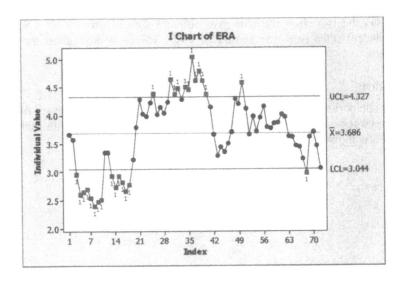

(c) The fact that many points are outside the control limits would seem to suggest that a different method be used in constructing the limits. However, the out-of-control points can be easily explained, as the early 1900s constituted the dead ball era; the 1930s was known as the hitters' era, and the late 1990s and early part of this century are becoming known as the steroid era. Although the left skewness in the entire set of data would suggest that a non-normal distribution be fit to the data and percentiles of that distribution be used to determine the control limits, there is not enough data to allow such an approach to produce reliable results.

(d) The *I*-chart is shown below, with the solid line just above the center line denoting the UCL (of 4.327) based on the first 72 observations. (The labeled UCL and LCL are computed from the 31 observations that are plotted and are shown here for comparison. None of the points plot below the LCL (not shown) for the 72 observations, and a point does plot above the UCL until 1987.

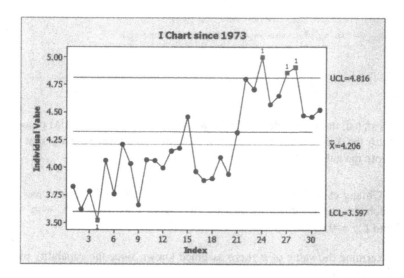

The CUSUM chart is given below, with the decision line of 5 shown. There was only one (slightly) negative value of S_L, with the other 30 values being zero. The first time that S_H exceeds 5 is 1983. Therefore, the signal is received four years before there is a signal on the *X*-chart (*I*-chart). Initial control chart limits should be computed from at least 100 observations; here 72 are being used but this is not a sample from the 72 years. Rather, this is 'the population" for those years so there is no sampling variability and the control limits thus should not be doubted on the basis of insufficient data, nor should the CUSUM calculations be doubted.

However, this is time series data so autocorrelation should be suspected and indeed it does exist here as the first five autocorrelations (i.e., lag one through lag five) are all large (.88, .78, .69, .60, and .51, respectively, and are above their respective decision limits. The control limits should be appropriately adjusted but that will not be done here because this was not covered in the text and is beyond the scope of the text.

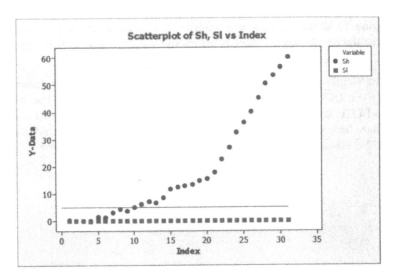

(e) As stated, the shift is detected four years sooner with the CUSUM procedure than with the X-chart, but a 'proper" estimate of sigma must be made that does not ignore the autocorrelation.

11.29. Chiang et al. [*Quality Engineering,* 14(2), 2001-2002) gave a case study for which the specifications for density (R1, g/cm³) are 1.14-1.15, with C_p = 0.67 and C_{pk} = 0.61.

(a) Determine the value of σ (here assumed known since the capability indices are not estimated).

(b) Using this value of σ, can μ be determined? Why or why not? If not, give two possible values of μ. Are these the only possible values? Explain.

Solution:

(a) $C_p = \dfrac{USL-LSL}{6\sigma}$ so $0.67 = \dfrac{1.15-1.14}{6\sigma}$ $\Rightarrow \sigma = .0025.$

(b) Since C_{pk} = 0.61, we can solve for μ using each of the two fractions (if necessary) in the definition of C_{pk} given in Section 11.15 since C_{pk} is defined as the minimum of the two values. Using the second fraction, we obtain 1.83 = $\dfrac{\mu - 1.14}{0.0025}$ $\Rightarrow \mu = 1.14458$. We can see by inspection that this value of μ results in the other fraction exceeding 1.83, so this must be the value of μ since C_{pk} is defined as 1/3 of the minimum of the two fractions.

11.31. If the LCL for an np-chart is 0.83, what must be the LCL for the corresponding p-chart if $n = 100$?

Solution:

0.83/100 = .0083

11.33. Twenty samples of size 200 are obtained and $\bar{p}$ = .035. What was the total number of nonconforming units for the 20 samples?

Solution:

The number is 20(200)(.035) = 140.

11.35.   A quality engineer computes $\widehat{C}_{pk}$ and finds that $\widehat{C}_{pk}$ = 1.02.   If the process characteristic were normally distributed and the specification limits were equidistant from $\overline{\overline{X}}$, the average of the subgroup averages, what percentage of units would be nonconforming (i.e., outside the specification limits)?

Solution:

$\widehat{C}_{pk}$ = 1.02 $\Rightarrow$ each Z = 3.06, 2(P(Z > 3.06)) = .0022. Therefore, 0.22% are nonconforming.

11.37.   Quality improvement results when variation is reduced. Assume a normal distribution with a known standard deviation, with 34% of the observations contained in the interval $\mu \pm a$. If the standard deviation is reduced by 50% and the mean is unchanged, what percentage of observations will lie in the interval $\mu \pm a$?

Solution:

For a normal distribution, 34% of the observations in the interval $\mu \pm a$ implies that the interval can be written as $\mu \pm 0.44\sigma$. For the interval to remain unchanged, a 50% reduction in σ means that the multiplier must change to .88. Then, from a standard normal table (or computer), the interval $\mu \pm 0.88\sigma^*$, with σ^* denoting the new standard deviation, must contain 62% of the observations.

11.39. Critique the following statement: 'Process capability must be good if control charts indicate that a company's processes are in a state of statistical control."

Solution:

Process capability and process control are essentially unrelated. A process can be in control but still not be a capable process, relative to the specification limits, because specification limits are not involved in the determination of control limits.

11.41. What are the two assumptions for an X-chart?

Solution:

The two assumptions for an X-chart are that the observations are independent and (approximately) normally distributed.

11.43. An $\overline{X}$-chart is to be constructed for Phase I. Given that number of subgroups = 24, subgroup size = 5, $\overline{\overline{X}} = 36.1$, and $\overline{s} = 6.2$, compute the 3-sigma control limits. If you had been given the data, what would you have done first before computing the limits?

Solution:

The 3-sigma control limits are: $\overline{\overline{X}} \pm 3 \dfrac{\frac{\overline{s}}{c_4}}{\sqrt{n}} = 36.1 \pm 3 \dfrac{\frac{6.2}{0.9400}}{\sqrt{5}} = 36.1 \pm 8.85$, so the control limits are $(27.25, 44.95)$. The first step should have been to check for approximate normality of the individual observations.

11.45. We know that mean shifts (and shifts in general) are detected much faster with a CUSUM procedure than with a Shewhart chart. Explain why that happens.

Solution:

A CUSUM procedure uses all of the available data from the time that the sums have been set or reset. This naturally includes data both before and after a parameter change. Since data after the change are thus used in the CUSUM computation, there is stronger evidence to suggest that a change has occurred than with a Shewhart chart, which would use out-of-control data only at a single point.

11.47. Assume that there has been a $1.6\sigma_x$ increase in the process mean. If X-chart limits were computed using known parameter values,

(a) How many points would we expect to have to plot on the X-chart before observing one that is above the UCL?

(b) If an $\overline{X}$-chart with subgroups of size $n = 4$ were used instead of an X-chart, how many points would we expect to observe before seeing a point above the UCL on the $\overline{X}$-chart?

(c) Compare your answers and explain why your answer to part (b) is smaller than your answer to part (a).

(d) Are your answers to (a) and (b) parameter-change ARLs or not? Explain.

Solution:

(a) For the UCL: $Z = \dfrac{(\mu + 3\sigma_x) - (\mu + 1.6\sigma_x)}{\sigma_x} = 1.4$ $P(Z > 1.4) = .0808$,

so the expected number of points before a point exceeds the UCL is $1/.0808 =$
12.383.

(b) For the UCL: $Z = \dfrac{(\mu + 3\frac{\sigma_x}{\sqrt{n}}) - (\mu + 1.6\sigma_x)}{\frac{\sigma_x}{\sqrt{n}}}$ with $n = 4$. Thus, $Z =$

$\dfrac{-0.1\sigma}{\frac{\sigma}{2}} = -0.2$ $P(Z > -0.2) = .579$ so the ARL $= 1/.579 = 1.726$.

(c) The answer in part (b) is (much) smaller than the answer in part (a) because
an $\overline{X}$-chart is more powerful than an X-chart in detecting mean shifts.

(d) Yes, these are both parameter-change ARLs because the mean has changed.

11.49. Explain why control chart parameters should not necessarily be
estimated using the same estimators for Phase I and Phase II.

Solution:

The objective should be to use the most appropriate estimator for each stage. In
Phase 1 it is desirable to use estimators that are not unduly influenced by bad
data points, whereas this is not a concern for Phase II since there are assumed to
be no bad data points when the stimation is done for Phase II, at the end of
Phase I. For the latter, the objective should be to use unbiased estimators that
have the smallest variance among all unbiased estimators that might be
considered.

11.51. Assume that a company is using an $\overline{X}$-chart with $n = 4$ to monitor a
process and is simultaneously using an X-chart because it is worried about the
possible occurrence of some extreme observations that might not be picked up by
the $\overline{X}$-chart. Assume that the control limits for each chart were constructed using
assumed values of $\mu = 20$ and $\sigma = 4$. If a point plots above the UCL on the $\overline{X}$-
chart, by how much does it have to exceed the UCL in order to be certain that at
least one of the four subgroup observations must plot above the UCL on the X-
chart?

Solution:

The UCL for the $\overline{X}$-chart is $20 + 3\dfrac{4}{\sqrt{4}} = 26$, and the UCL for the X-chart is $20 +$

$3(4) = 32$. Since all four observations in the subgroup could theoretically be the
same, the subgroup average would have to lie more than 6 units above its UCL
to ensure that at least one observation would exceed the UCL on the X-chart.

11.53. As discussed in Section 11.9.1, the use of runs rules is inadvisable. Explain why the in-control ARL of an $\overline{X}$-chart with even one run rule (such as 8 points above the midline) must be less than the in-control ARL without the addition of the run rule?

Solution:

This is because any additional criterion can only reduce the in-control ARL since a signal can then be received from a second source.

11.55. A combined Shewhart-CUSUM scheme is used for averages with 3.5-sigma limits and $h = 5$ and $k = 0.5$ for the CUSUM scheme. At what point will a signal be received if five consecutive subgroup averages transform to Z-values of 3.4, 1.86, 1.25, 0.91, and 0.55? Explain. Although a signal was received, what does the pattern of $\overline{X}$ values suggest to you? In particular, what course of action would you take as the process engineer, if any?

Solution:

A signal is received on the third subgroup as $S_H = 5.01$. The fact that the $\overline{X}$ values are tending toward what would be the midline on the $\overline{X}$-chart suggests that the likely out-of-control condition is being rectified.

11.57. Consider the control chart given at

http://www.minitab.com/resources/stories/ExhibitAPg3.aspx that was described in Section 11.1. Can we tell from the graph if the points in red were included or not when the control limits were computed? Does the text offer any hint?

Solution:

It is difficult to tell from the graph. The sum of the signed distances from the points to the line should be zero. If those points were included, the sum would seem to be positive but that may be just an illusion caused by the fact that although the extreme points that are above the midline are a greater distance from it than the extreme points that are below it, there are more points below the midline than above it. The text suggests that all points were probably used in the computations. It would have been better to exclude the points but including them and still having the points above the UCL of course gave the homeowners a stronger argument.

11.59. Moving average charts were mentioned briefly, but not advocated, in Section 11.5. Compute the correlation between consecutive moving averages of (a) size 3, (b) size 5, and (c) size n. What would be one advantage of using a moving average chart with averages computed from at least a moderate number of observations (say, at least 5)? If a user insists on using a moving average chart, what would be your recommendation?

Solution:

(a) $n = 3$:

The covariance is $2\sigma^2/9$ since moving averages of size 3 have two values in common. Thus, the correlation is:

$$\rho = \frac{2\sigma^2/9}{\sqrt{\sigma^2/3}\,\sqrt{\sigma^2/3}} = 2/3$$

(b) $n = 5$:

The moving averages have four observations in common, so the correlation is:

$$\rho = \frac{4\sigma^2/25}{\sqrt{\sigma^2/5}\,\sqrt{\sigma^2/5}} = 4/5$$

(c) Similarly, for moving averages of size n the correlation can be shown to be $(n - 1)/n$. Moving averages smooth out the data, with the amount of smoothing dependent upon the number of observations used in each average. In process control applications we generally do not want to smooth out the data, as we want to detect out-of-control conditions. In some applications, however (such as described in the next problem) there are intermittent blips that are considered by process engineers to be a (relatively) normal part of the process. There would thus be no reaction to such blips so a statistic that hides them might be used.

I would recommend that moving average charts generally not be used in other applications, however.

11.61. An application of a *p*-chart is given in the *NIST/SEMATECH e-Handbook of Statistical Methods* at
http://www.itl.nist.gov/div898/handbook/pmc/section3/pmc332.htm.
Specifically, chips on a wafer were investigated and a nonconforming unit was said to occur whenever there was a misregistration in terms of horizontal and/or vertical distances from the center. Thirty wafers were examined and there were 50 chips on each wafer, with the proportion of misregistrations over the 50 chips per wafer recorded for each wafer. The average of the proportions is .2313.

(a) Would it be necessary to use the Ryan-Schwertman control limits in Section 11.8.1.1, or should 3-sigma limits suffice? Explain.

(b) Compute both sets of limits and comment.

Solution:

(a) Since $\bar{p}$ = .2313, the binomial distribution, if appropriate, is definitely skewed, so it would be better to use the Ryan-Schwertman limits than to use symmetric limits.

(b) The 3-sigma limits are:

$$.2313 \pm 3\sqrt{\frac{(.2313)(.7687)}{50}} = .2313 \pm .1789 = (.0524, .4102)$$

The Ryan-Schwertman limits are:

$$\text{UCL} = \frac{1}{n}(0.6195 + 1.00523np + 2.983\sqrt{np})$$

$$= \frac{1}{50}(0.6195 + 1.00523(50)(.2313) + 2.983\sqrt{50(.2313)}\,)$$

$$= .4492$$

$$\text{LCL} = \frac{1}{n}(2.9529 + 1.0195np - 3.2729\sqrt{np})$$

$$= \frac{1}{50}(2.9529 + 1.0195(50)(.2313) - 3.2729\sqrt{50(.2313)}\,)$$

$$= .0723$$

Thus, the Ryan-Schwertman limits differ noticeably from the 3-sigma limits, although the difference isn't great.

11.63. One of the sample datasets that comes with the MINITAB software is PIPE.MTW. A company that makes plastic pipes is concerned about the consistency of the diameters and collects the following diameter measurements over a 3-week period.

Week 1	Week 2	Week 3	Machine	Operator
5.19	5.57	8.73	1	A
5.53	5.11	5.01	2	B
4.78	5.76	7.59	1	A
5.44	5.65	4.73	2	B
4.47	4.99	4.93	1	A
4.78	5.25	5.19	2	A
4.26	7.00	6.77	1	B
5.70	5.20	5.66	2	B
4.40	5.30	6.48	1	A
5.64	4.91	5.20	2	B

Construct the appropriate control chart to determine if the diameter measurements are in control. Does the chart suggest control? If not, is there evidence of a lack of statistical control relative to either machine, or either operator, or a particular machine-operator combination?

Solution:

The individual observations chart is given below. Although the point that is above the UCL corresponds to Operator A using Machine 1, there appears to be a time effect as the sums of the observations are increasing more than slightly over the three weeks and the point above the UCL was in Week 3.

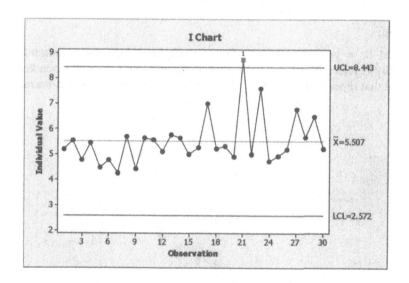

11.65. The sample dataset PAPER.MTW that comes with the MINITAB software contains measurements on rolls of paper used by a large metropolitan newspaper. The engineer who inspects each roll of paper is especially concerned about small holes in the paper that are larger than 1 millimeter in diameter. The dataset contains 35 1.5 x 2 sections of paper, with each being sampled 6 times. There are thus 210 numbers, each of which represents the number of holes of at least 1 millimeter for the given section. Would it be reasonable to use the 210 counts to determine Phase I control limits for a c-chart? If so, what assumption(s) would you be making? Determine the control limits, using whatever method seems best in view of the small number of counts. Then test for control of the 210 measurements.

Solution:

The 3-sigma UCL is 3.904 and there is no LCL since $\bar{c}$ is (much) less than 9. Four points plot above the UCL. The regression-based limits are UCL = 4.58 and LCL = 0.74. Seventy-seven points plot below this LCL and three points plot above it. Since zero is the best possible count, we would hope that zero plots below a LCL. In determining these limits we are assuming independence within each section of paper and between sections. These assumptions should preferably be checked.

11.67. Consider the $\bar{X}$-chart in Figure 11.6. Even though subgroup averages are plotted rather than individual observations, should a normal probability plot have been constructed before the $\bar{X}$-chart was constructed? Explain.

Solution:

It would be a good idea to test for normality and here the assumption of normality is rejected with a normal probability plot, with the plot given below. (We see that repeated values along with skewness have caused the non-normality signal.)

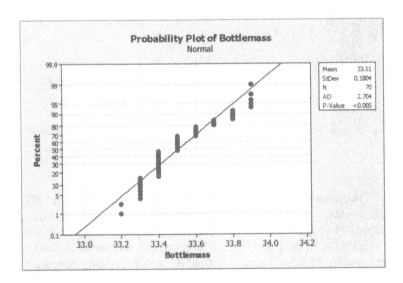

When this occurs a judgment decision must be made as to whether or not the subgroup size of 5 is enough to compensate for the degree of non-normality that is present. Here a dotplot of the observations shows right skewness so the $\bar{X}$-chart with 3-sigma limits is a bit shaky. (There is not much point in constructing any type of plot of subgroup averages because there are only 14 of them. Such a small number will not give us any meaningful distributional information.)

12

Design and Analysis of Experiments

12.1. Use the following applet, due to Christine Anderson-Cook, to do a one-factor two-level test (left hand versus right hand):
http://www.amstat.org/publications/jse/java/v9n1/anderson-cook/BadExpDesignApplet.html.

Solution:
(student Internet exercise)

12.3. Assume that data from a 2^3 design have been analyzed, and one or more of the interactions are significant. What action should be taken in investigating main effects for factors that comprise those interactions?

Solution:
The conditional effects should be computed.

12.5. Explain why residual plots should be used with data from designed experiments.

Solution:
Residual plots can be used in various ways, including checking the equal variance assumption for replicated designs. In general, residual plots should be used in essentially the same way with Analysis of Variance models that they are used with regression models.

12.7. Oftentimes the design of an experiment is lost and all that is available are the treatment combinations and the values of the response variable. Assume that four factors, each at two levels, are studied with the design points given by the following treatment combinations: (1), ab, bc, abd, acd, bcd, d, and ac.

(a) What is the defining contrast?

(b) Could the design be improved, using the same number of design points?

(c) In particular, which three main effects are confounded with two-factor interactions?

Solution:

(a) Since eight design points were used, which is half the number that would be used with a full factorial, the design is a half fraction with a single defining contrast. It is easy to see that the latter is $I = ABC$ because all of the design points have an even number of letters in common with ABC.

(b) Half fractions should be constructed by using the highest-order interaction as the defining contrast, as this will maximize the resolution of the design. Therefore, $I = ABCD$ would have been a better choice, in general.

(c) A, B, and C are each confounded with two-factor interactions.

12.9. Assume that ANOVA is applied to a single factor with four levels and the F-test is significant. What graphical aid could be used to determine which groups differ in terms of their means?

Solution:
Analysis of Means (ANOM) might be used.

12.11. An experimenter decides to use a 2^{6-2} design and elects to confound ABD and CEF in constructing the fraction.

(a) Determine what effect(s) would be aliased with the two-factor interaction AB.

(b) Is there a better choice for the two defining contrasts?

Solution:
(a) $AB = D = ABCEF = CDEF$

(b) A better choice would be $I = ABCD = CDEF$. Such a design would not confound main effects with two-factor interactions, which is unnecessary.

12.13. Assume that a 2^4 design is to be run. Explain to the person who will conduct the experiment what the treatment combination ab means.

Solution:
This means to set factors A and B at their high levels and factors C and D at their low levels.

12.15. How many effects can be estimated when an unreplicated 2^3 design is used, and what are those effects? (List them.)

Solution:
All seven effects are estimable. They are: A, B, C, AB, AC, BC, and ABC.

12.17. Assume a one-factor experiment with the factor being random. Explain, *in words*, the null hypothesis that is tested.

Solution:

In words, what is tested is that the variance of the response due to the treatment effect is zero over the entire range of possible levels of interest, which is the same as saying there is no treatment effect and that the variation in the response is simply random variation.

12.19. When Analysis of Means is applied to certain full or fractional factorial two-level designs (unreplicated), the UDL will be above the midline on the chart by an amount equal to $t_{\alpha/2,\nu}\, s/4$. Name one of the designs for which this is true.

Solution:

The general expression for the distance above the midline when two means are compared is $t_{\alpha/2,\nu}\frac{s}{\sqrt{2n}}$. Here $\sqrt{2n} = 4$, so $n = 8$. A 2^4 design would have each average computed from 8 numbers, so that would be one possible design.

12.21. Assume that Analysis of Means is being applied to data from a one-way classification (i.e., one factor) with five levels. State the general form of one of the five hypothesis tests that would be performed.

Solution:

Each test would test the null hypothesis that the mean response at a given level is equal to the average of the mean responses, averaged over all five levels.

12.23. The following statement is paraphrased from an Internet message board: "I have two factors, each at two levels. When I multiply the factor levels together to create the levels for the interaction, the interaction levels are $+1$ and -1, but when I look at the combinations of the factors in the original units I see four combinations. How can $+1$ and -1 each represent two different combinations of factor levels?" Explain the faulty reasoning in these statements.

Solution:

In raw form, it is better to say that the product of factor levels constitutes a product, not an interaction, since the product will have more "levels" than will each of the factors. In experimental design, one is interested in looking at interactions, not products, which is why the analysis must be performed in coded form.

12.25. The data given in ENGSTAT1224.MTW are from a statement by Texaco, Inc. to the Air and Pollution Subcommittee of the Senate Public Works Committee on June 26, 1973. It was stated that an automobile filter developed by Associated Octel Company was effective in reducing pollution, but questions were raised in regard to various matters, including noise. The president of Texaco, John McKinley, asserted that the Octel filter was at least equal to standard silencers in terms of silencing properties. The data in the file are in the form of a 3×2^2 design with SIZE of the vehicle being 1 = small, 2 = medium, and 3 = large; TYPE being 1= standard silencer and 2 = Octel filter, and SIDE being 1= right side and 2 = left side of the car. The response variable is NOISE.

(a) Are these factors fixed or random, or can that be determined? (Remember that the classification of a factor as fixed or random affects the way the analysis is performed when there is more than one factor.)

(b) Since a 3×2^2 design (a mixed factorial) was not covered explicitly in the chapter, initially ignore the SIZE factor and analyze the data as if a replicated 2^2 design had been used, bearing in mind that the objective of the experiment is to see if the Octel filter is better than the standard filter.

(c) Now analyze the data using all three factors, using appropriate software such as the general linear model (GLM) capability in MINITAB, and compare the results. In particular, why is one particular effect significant in this analysis that was not significant when the SIZE factor was ignored? (*Hint*: Look at the R^2 values ignoring and not ignoring SIZE.)

(d) Is there an interaction that complicates the analysis? If so, what approach would you take to explain the comparison between the standard filter and the Octel filter?

Solution:

(a) The factors are fixed. The standard silencer may have been selected at random from various makes of standard silencers, but the Octel filter was obviously not selected at random.

(b) Analyzing the data as a replicated 2^2 design shows no effect to be significant. Remarkably, the R^2 value for the model with both main effects and the interaction effect is only .036. We would expect to see a larger value of R^2 due to chance alone when a model contains three terms.

(c) The analysis is given below. Notice that the TYPE factor is significant here but was not significant when SIZE was initially ignored. When terms that should be in a model are not included, the error term will be inflated, which can cause real effects to not be declared significant.

(d) SIZE and TYPE are now clearly significant, but a conditional effects analysis would have to be performed in assessing the effects of each factor because of the large SIZE x TYPE interaction.

```
General Linear Model: NOISE versus SIZE, TYPE, SIDE

          Factor  Type   Levels  Values
          SIZE    fixed       3  1, 2, 3
          TYPE    fixed       2  1, 2
          SIDE    fixed       2  1, 2


Analysis of Variance for NOISE, using Adjusted SS for Tests
```

```
Source      DF    Seq SS    Adj SS    Adj MS      F       P
SIZE         2   26051.4   26051.4   13025.7   893.19  0.000
TYPE         1    1056.3    1056.3    1056.3    72.43   0.000
SIDE         1       0.7       0.7       0.7     0.05   0.829
SIZE*TYPE    2     804.2     804.2     402.1    27.57   0.000
SIZE*SIDE    2    1293.1    1293.1     646.5    44.33   0.000
TYPE*SIDE    1      17.4      17.4      17.4     1.19   0.286
SIZE*TYPE*SIDE 2  301.4     301.4     150.7    10.33   0.001
Error       24     350.0     350.0      14.6
Total       35   29874.3
```

$$S = 3.81881 \quad R-Sq = 98.83\% \quad R-Sq(adj) = 98.29\%$$

12.27. If you are proficient in writing MINITAB macros, write a macro that could be used for a 2^3 design such that the main effects and interaction effects are all shown on the same display, with the line that connects the points for each interaction computed as discussed in Section 12.7. Then use your macro to construct an ANOM display for the data in Section 9.3, using the center-point runs to estimate σ_ϵ.

Solution:
(student programming exercise)

12.29. When should a randomized block design be used? If an experiment were run and the block totals were very close, what would that suggest?

Solution:
A randomized block design should be used when an experimenter believes that the experimental units are not homogeneous. If such a design were used and the block totals were very close, this would suggest that blocking probably wasn't necessary.

12.31. Construct a standard OFAT design for five factors and compute the correlations between the columns of the design. Do you consider the latter to be unacceptably large? Why or why not?

Solution:
The design is given below. Six of the correlations are .111 and the other four are .236. Thus, the correlations are not especially large.

```
Row    A    B    C    D    E
 1    -1   -1   -1   -1   -1
 2    -1    1    1    1    1
 3     1   -1    1    1    1
 4     1    1   -1    1    1
 5     1    1    1   -1    1
 6     1    1    1    1   -1
 7     1    1    1    1    1
 8     1   -1   -1   -1   -1
```

9	-1	1	-1	-1	-1
10	-1	-1	1	-1	-1
11	-1	-1	-1	1	-1
12	-1	-1	-1	-1	1
13	1	1	-1	-1	-1
14	-1	1	1	-1	-1
15	-1	-1	1	1	-1
16	1	-1	1	-1	-1
17	1	-1	-1	1	-1
18	-1	1	-1	1	-1

12.33. Usher and Srinivasan [*Quality Engineering*, 13(2), 161-168, 2000-2001] described a 2^4 experiment for which there were six observations per treatment combination, but the data were not analyzed as having come from a replicated experiment. Specifically, the authors stated: "Note that these six observations do not represent true replicates in the usual sense, because they were not manufactured on unique runs of the process." Consequently, the average of the six values at each treatment combination was used as the response value, and the data were thus analyzed as if they had come from an unreplicated experiment. Explain how analyzing the data as if the data had come from a replicated experiment could produce erroneous results.

Solution:
It is important to note the distinction between multiple readings and replications. The former will generally be correlated because they are usually made close together in time. Replications result when an experiment is repeated (factors are re-set, etc.). Using multiple readings as if they were replicates will likely underestimate the standard deviation of the error term in the model and cause factors to be erroneously declared significant.

12.35. A. Gupta (*Quality Engineering*, 1997-98) presented a case study involving antibiotic suspension products, with "separated clear volume" (the smaller the better) being the response variable. The objective was to determine the best level of each of five two-level factors so as to minimize the response variable, with the selection of a particular level of a factor being crucial only if the factor is significant. An L_{16} orthogonal array [see part (b)] was used, and that design and the accompanying response values are as follows:

Response	A	B	C	D	E
47	8	50	0.2	0.4	Usual
42	8	50	0.4	0.4	Modified
50	8	60	0.4	0.4	Usual
51	8	60	0.2	0.4	Modified
40	16	50	0.4	0.4	Usual
44	16	50	0.2	0.4	Modified
46	16	60	0.2	0.4	Usual
40	16	60	0.4	0.4	Modified
28	8	50	0.2	0.6	Usual
23	8	50	0.4	0.6	Modified
30	8	60	0.4	0.6	Usual
38	8	60	0.2	0.6	Modified
26	16	50	0.4	0.6	Usual
31	16	50	0.2	0.6	Modified
32	16	60	0.2	0.6	Usual
33	16	60	0.4	0.6	Modified

(a) Should the data be analyzed in the original units of the factors (with of course some numerical designation for the two levels of factor E)? Why or why not?

(b) Show that this L_{16} array is equivalent to a suboptimal 2^{5-1} design. (*Hint*: The interaction that is confounded with the mean can be determined using MINITAB, for example, by trying to estimate all the effects. That is, by using a command such as FFACT C1= (C2-C6)5, with the response values being in C1 and the factors being in C2-C6.

(c) Explain the consequences of using this L_{16} array rather than the 2^{5-1} design with maximal resolution in terms of the estimation of the two-factor interactions.

(d) Estimate as many effects as you can with this design and determine the significant effects. With the second level (60) of the second factor considered to be preferable for nonstatistical reasons, what combination of factor levels appears to be best? Is it necessary to qualify your answer in any way since the L_{16} design is a resolution IV design whereas a 16-point resolution V design could have been constructed?

Solution:
(a) The data should be analyzed using coded factor levels. If the true model did not contain interaction terms, then using the actual levels would be okay, but we never know the true model and we cannot assume that there won't be any interactions. The presence of interactions creates problems when the data are analyzed using the actual levels.

(b) The defining relation is $I = ABCE$, whereas $I = ABCDE$ is needed for a resolution V design.

(c) Certain two-factor interactions will be confounded with other two-factor interactions with this resolution IV design, whereas all two-factor interactions would have been confounded with three-factor interactions if the design had been constructed as a resolution V design.

(d) A normal probability plot shows the D effect to be the only significant effect. That effect is confounded with a 5-factor interaction, so there is no interpretation problem for that effect. Since the Pareto chart (see below) shows the second factor to not quite be significant, it shouldn't make any difference which level is used. The same can be said of the other factors except factor D, the high level of which is clearly preferable for minimizing the response. The fact that the design does not have the maximum resolution is not a problem in this application because of the paucity of significant effects.

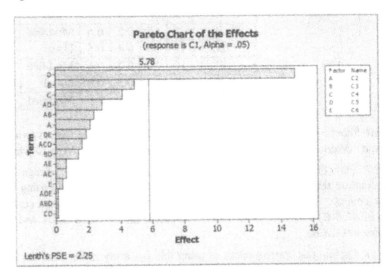

12.37. The following interaction profile shows the results of an unreplicated 2^2 design.

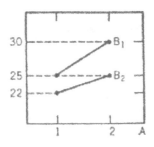

Estimate the A effect.

Solution:

The estimate of the A effect is $\overline{A}_{high} - \overline{A}_{low} = \dfrac{30+25}{2} - \dfrac{25+22}{2} = 4.0$

12.39. L. B. Hare (*Journal of Quality Technology*, Vol. 20, pp. 36-42, 1988) reported the use of a 2^{5-1} design in an experiment for which the primary goal was to determine the effect of five factors on fill weight variation for a dry mix soup filling process. The largest effect estimate was the main effect estimate of factor E (3.76), and the largest two-factor interaction involving this factor was the BE interaction, for which the effect estimate was 3.24. This was the second largest effect.

(a) How would you recommend that the influence of factor E be reported?

(b) What effect is the E effect confounded with, assuming that the design was constructed in an optimal manner? Should this confounding be of any concern?

(c) Similarly, what effect is the BE interaction confounded with and should that confounding be of any concern?

Solution:

(a) Since the BE interaction is large relative to the E effect, the conditional effects of E should be reported, as they will differ considerably.

(b) The E effect is confounded with the $ABCD$ interaction, which should be of no concern since four-factor interactions are very rarely real effects.

(c) The BE interaction is confounded with the ACD interaction, which could be significant, although three-factor interactions generally are not real effects.

12.41. Assume that Analysis of Variance calculations have been performed for a problem with two levels of a single factor. This would produce results equivalent to an independent sample t-test provided that the alternative hypothesis for the t-test is (choose one): (a) greater than, (b) not equal to, or (c) less than.

Solution:
(b) not equal to

12.43. Analyze the following data from a 2^2 design with three replicates.

		A Low	A High
B	Low	10 12 16	8 10 13
	High	14 12 15	12 15 16

Solution:
None of the effects are significant as the smallest p-value is .104. It should be noted, however, that conditional main effects for factor B, in particular, should be reported because of the relatively large interaction.

12.45. A 2^2 design with two replicates is run and the following results are obtained. The response totals are: 20 when both factors are at the low level, 30 when both factors are at the high level, 15 when factor A is at the high level and factor B is at the low level, and 18 when factor A is at the low level and factor B is at the high level. Determine the estimate of the A effect.

Solution:
The estimate of the A effect is $\dfrac{15+30}{2} - \dfrac{20+18}{2} = 3.5$.

12.47. Consider a 2^2 design. Construct an interaction plot for which the estimate of the interaction effect is zero, being careful to do the appropriate labeling.

Solution:
Any interaction plot in the form of an X will suffice, as will a plot in which the lines that form the X are at more acute or less acute angles, but form a symmetric figure.

12.49. An unreplicated 2^{5-1} design is used and a normal probability plot is subsequently used. If all effects that can be estimated are used for the plot:

(a) How many points will the plot contain?

(b) What would you suspect if the plot showed the vast majority of the effects to be significant?

Solution:
(a) 15

(b) I would suspect at least one bad data point.

12.51. The issue of analyzing data from a designed experiment in coded form versus raw form was covered and illustrated numerically in Section 12.12. Consider a model with two main effects and an interaction term.

(a) Show algebraically why the main effect model coefficients will be different for the two forms, and thus the p-values will differ.

(b) Use 12, 10, 21, and 28, respectively, and 23, 26, 18, and 14, respectively, as the two sets of response values for the treatment combinations in the example in Section 12.12 with the two sets of values being replicates of the treatment combinations used and show numerically that the p-values for the two main effects differ between the raw-form data and the coded-form data.

Solution:
(a) This was illustrated in Section 13.11 of the 2nd edition of *Statistical Methods for Quality Improvement* by T. P. Ryan, and is also illustrated in Example 4.1 of *Modern Experimental Design* by the same author. In coded form, the terms in the model are A, B, and AB. The discrepancy results from the fact that the raw-form

terms are not A^*, B^*, and A^*B^*. Specifically, if we multiply A times B, we create some additional terms.

That is, $\left(\dfrac{A^*-A_{ave}^*}{half-distance\ between\ levels}\right)\left(\dfrac{B^*-B_{ave}^*}{half-distance\ between\ levels}\right)$

will when multiplied out create linear terms in A^* and B^* in addition to the terms that represent the linear (main) effects. Thus, the p-values for main effects differ between the two models because the models are not equal in terms of main effects.

(b) The p-value is .938 for the interaction term in both model forms. In coded form, the p-values for A and B are .876 and .700, respectively, whereas the raw-form p-values are .963 and .961, respectively.

12.53. Explain, in words, what is being tested when an experiment has two fixed factors and their possible significance is to be tested with hypothesis tests.

Solution:
For each factor, one is testing whether there is a significant difference between the average response values at each of the levels of the factor.

12.55. Construct an example for a single (fixed) factor with three levels and five observations per level for which the overall F-test shows a significant result but the averages for the three levels are 18.1, 18.2, and 18.3, respectively, and all 15 numbers are different.

Solution:

Three levels with 5 observations per level:

18.08	18.18	18.28
18.09	18.19	18.29
18.10	18.20	18.30
18.11	18.21	18.31
18.12	18.22	18.31

The column averages are obviously 18.10, 18.20, and 18.30, and the F-test for testing the equality of the means results in an F-statistic of 219.43.

12.57. Sopadang, Cho, and Leonard [*Quality Engineering*, 14(2), 317-327, 2001] described an experiment in which 15 factors were examined in 16 experimental runs. What was the resolution of the design and for what purpose would you suggest that the design be used?

Solution:
It would have to be a resolution III design as only main effects would be estimable. Such a design would generally be used as a screening design.

12.59. There are eight treatment combinations used when a 2^{5-1} design is run with $I = ABCDE$. Which one of the following treatment combinations is used in error with the other four (correct) treatment combinations: a, bcd, ac, bce, ace?

Solution:
The treatment combination that is out of place is *ac*. One easy way to see this is to notice that this treatment combination has an even number of letters in common with the interaction that constitutes the defining contrast, whereas the other four treatment combinations have an odd number of letters in common with it.

12.61. Explain to a person unfamiliar with experimental design what (1) denotes when used to represent a treatment combination and explain why it is used.

Solution:
The symbol (1) is used to represent the low level of each factor that is used in an experiment.

12.63. Explain why a person who runs an experiment using a 2^3 design does not want to see a significant *ABC* interaction effect.

Solution:
A significant *ABC* interaction would mean that the conditional effects for the two-factor interactions would differ considerably and thus the usual (single) effect estimate of each would be unreliable.

12.65. Construct an example of data from a 2^2 design for which the estimate of the *A* effect is 0 and the estimate of the *B* effect is 5.

Solution:
This could be accomplished with the treatment combinations 10, 10, 5, and 15, the first two numbers being at the low level of *A*.

12.67. What course of action would you recommend if two interaction effects for a 2^4 design exceeded two of the main effects for factors that comprise the interaction effects?

Solution:
The corresponding conditional (half) effects should be computed.

12.69. Explain the condition(s) under which the selection of a 1/4 fraction of a full factorial design should not be chosen randomly.

Solution:
Hard-to-change factors and impossible or impractical factor-level combinations often occur and can obviate the random choice of a fraction. Instead, it may be necessary to select a fraction in which all of the treatment combinations are feasible.

12.71. To see the importance of maintaining processes in a state of statistical control when experiments are performed, consider the following. An industrial experiment is conducted with temperature set at two levels, 350 °F and 450 °F.

Assume that no attempt at process control is made during the experiment and consequently there is a 3-sigma increase in the average conductivity from the original value of 13.2, due to some cause other than the change in temperature, and the increase occurs right when the temperature is changed. Assume that σ is known to be 4.5 and that 20 observations were made at each of the two temperature levels.

(a) What is the expected value of the appropriate test statistic for this scenario and the corresponding expected p-value?

(b) Answer these same two questions if the process went out of control, by the same amount, when the first 10 observations had been made at the second temperature. What argument do these two sets of numbers support?

Solution:

(a) $E(\bar{x}_1) = \mu_1$ $E(\bar{x}_2) = \mu_2 + 3\sigma$ with $\mu_1 = \mu_2$

Therefore, $E(Z) = \dfrac{-3\sigma}{\sigma\sqrt{\frac{1}{20} + \frac{1}{20}}} = -9.49.$ The associated p-value is essentially zero.

(b) $E(\bar{x}_2) = \mu_2 + \dfrac{3\sigma}{2}$ so $E(Z) = \dfrac{-3\sigma/2}{\sigma\sqrt{\frac{1}{20} + \frac{1}{20}}} = -4.74$

The associated p-value is zero to the first five decimal places. Both numbers show the importance of maintaining processes in control when experimentation is performed.

12.73. An experiment is run using a 2^3 design. The eight response values are as follows.

	C_{low}		C_{high}	
	B_{low}	B_{high}	B_{low}	B_{high}
A_{low}	16	18	22	24
A_{high}	19	14	21	27

(a) Construct the BC interaction plot.

(b) Based solely on this plot, would you recommend that the B and C main effects be reported? Why or why not?

Solution:

(a) The interaction plot is given below.

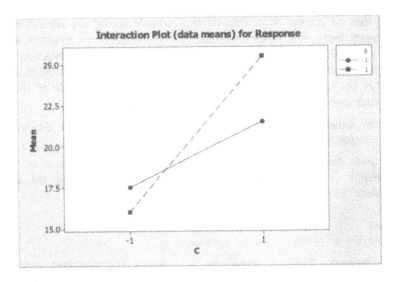

(b) The main effects for B and C should not be reported because the interaction plot shows a large interaction (since the lines cross)

12.75. If you have access to MINITAB, show that a Taguchi orthogonal array design constructed for seven factors, each at two levels, with eight experimental runs (i.e., the L_8 in Taguchi's notation), is the same design as the 2^{7-4} design produced by MINITAB. Comment.

Solution:
As is know now well known, virtually all of the designs that were previously known as "Taguchi designs" really aren't new designs at all; they just had a new name.

12.77. Assume that there is one quantitative variable that is believed to relate to the response variable and four qualitative variables that are arranged in the form of an experimental design. How would you suggest that the data be analyzed?

Solution:
This is a classic example of when Analysis of Covariance should be used so as to adjust for the effect of the quantitative variable, assuming that it is essentially a nuisance variable, at least relative to the design.

12.79. Consider the MINITAB output at the beginning of Section 12.9.2. How would you explain to a manager who has never studied experimental design what $B + ACE + CDF + ABDEF$ means.

Solution:

When a small number of experimental runs is made, relative to the number of factors used in the experiment, it won't be possible to estimate all of the effects, including all of the interactions between factors. Consequently, some effects will be intertwined, and it is customary to indicate these in the appropriate groups.

12.81. Consider the article referenced in Exercise 12.76. Can the main effect of the first factor, S, be interpreted unambiguously for that dataset? Explain.

Solution:
The predictors are correlated in that dataset, and for correlated predictors there is no such thing as an "effect estimate".

12.83. A design with five factors, each at two levels, was run and the alias structure involving factor E was $E = ABCD$. What is the resolution of the design?

Solution:
The design is resolution V because $I = ABCDE$.

12.85. Given the data in Example 12.6, verify the statement that there are no significant effects other than the block effect.

Solution:
The ANOVA table below shows that the main effects collectively are not even remotely close to being significant, and the same can be said of the 2-factor interactions. (As the problem states, the blocking was performed on the 3-factor interaction.) The block effect is highly significant.

```
                  Analysis of Variance
Source           DF  Seq SS  Adj SS  Adj MS   F      P

Blocks            3 152.563 152.562 50.8542 77.49  0.000
Main Effects      3   1.250   1.250  0.4167  0.63  0.619
2-Way Ints.       3   2.188   2.188  0.7292  1.11  0.415
Residual Error    6   3.937   3.937  0.6562
Total            15 159.938
```

13

Measurement System Appraisal

13.1. If a negative estimate of a variance component means that the corresponding factor or interaction has no effect, does a positive estimate of a variance component likely mean that the corresponding effect is significant? Explain.

Solution:
No, negative variance component estimates occur when F-statistics are small, such as when the F-statistic is less than one in one-factor ANOVA. F-statistics can be just large enough to result in positive variance component estimates, but be well short of the appropriate critical F-values.

13.3. A company performs a study to see if there is a machine effect that is contributing to its problems with excessive variability in a particular product line. Three machines are picked at random for the study and this is the only factor that is examined in the study. If $SS_{machines} = 72.4$, $MS_{error} = 12.4$, and there are 10 measurements made for each machine, what is the numerical value of $\hat{\sigma}^2_{machines}$?

Solution:
Since the machines are random, $E(MS_{machines}) = \sigma^2 + 10\sigma^2_{machines}$, so $\hat{\sigma}^2_{machines}$
$= \frac{MS_{machines} - \hat{\sigma}^2}{10}$. $SS_{machines} = 72.4$ so $MS_{machines} = 72.4/2 = 36.2$. $\hat{\sigma}^2 =$
$MS_{error} = 12.4$ so $\hat{\sigma}^2_{machines} = \frac{36.2 - 12.4}{10} = 2.38$

13.5. A useful class experiment would be the one mentioned at the beginning of the chapter. The students in a class (of at least moderate size) could each measure the width of the desk used by the instructor, using both a ruler or other measuring instrument that is at least as wide as the desk and one that is not as wide as the desk. Of course repeatability could be assessed by having each student make two or three measurements using the same measuring instrument.

Solution:
(classroom exercise)

13.7. Assume, relative to Exercise 13.6, that a foreperson states that the three operators involved in the study are the only ones who will be doing this work. Does the model given in Eq. (13.1) still apply? Explain.

Solution:
The model no longer applies as there is no variance component for a factor when the levels of the factor are fixed.

14

Reliability Analysis and Life Testing

14.1. Derive the reliability function for the Weibull distribution that was given in Section 14.5.3.

Solution:
$R(t) = 1 - F(t)$ and $f(t) = (\beta/\alpha)(t/\alpha)^{\beta-1}exp[-(t/\alpha)^\beta]$, so
$F(t) = \int_0^t (\beta/\alpha)(t/\alpha)^{\beta-1}exp[-(t/\alpha)^\beta]\, dt$. Let $u = (t/\alpha)^\beta$. Then
$R(t) = 1 - F(t) = 1 - (1 - exp[-(t/\alpha)^\beta])$, so $R(t) = exp[-(t/\alpha)^\beta]$

14.3. Verify Eq. (14.5).

Solution:
Eq. (14.3) is $\mu(S_1) = log[a(S_2, S_1)] + \mu(S_2)$.
Eq. (14.4) is $\mu(S) = log(R_0) + \dfrac{E_a}{k(S+273)}$

Thus, $log[a(S_2, S_1)] = \mu(S_1) - \mu(S_2) = log(R_0) + \dfrac{E_a}{k(S_1 + 273)} - log(R_0) -$

$\dfrac{E_a}{k(S_2 + 273)} = \dfrac{E_a}{k}[\dfrac{1}{(S_1+273)} - \dfrac{1}{(S_2+273)}]$ so

$a(S_2, S_1) = exp\{\dfrac{E_a}{k}[\dfrac{1}{(S_1+273)} - \dfrac{1}{(S_2+273)}]\}$.

14.5. Suppose that the time to failure, in minutes, of a particular electronic component subjected to continuous vibrations can be approximated by a Weibull distribution with $\alpha = 50$ and $\beta = 0.40$. What is the probability that such a component will fail in less than 5 hours?

Solution:
From the solution to Exercise 14.1,
$$F(t) = 1 - exp[-(t/\alpha)^\beta]$$
$$= 1 - exp[-(t/50)^{0.40}]$$
$$= 1 - exp[-(5/50)^{0.40}] = .328$$

14.7. It is known that the life length of a certain electronic device can be modeled by an exponential distribution with $\lambda = .00083$. Using this distribution, the reliability of the device for a 100-hour period of operation is .92. How many hours of operation would correspond to a reliability of .95?

Solution:
$R(100) = exp(-0.00083(100)) = .92$
$R(t) = .95 \Rightarrow exp(-0.00083t) = .95 \Rightarrow -0.00083t = log(.95)$, so $t = -\frac{log(.95)}{0.00083} = 62$ (rounded to the nearest integer).

14.9. Assume that an electronic assembly consists of three identical tubes in a series system and the tubes are believed to function (and fail) independently. It is believed that the function $f(t) = 50t\ exp(-25t^2)$ adequately represents the time to failure for each tube. Determine $R_s(t)$. Looking at your answer, does $f(t)$ appear to be a realistic function? Explain.

Solution:
$f(t) = 50t\ exp[-25t^2]$. Let $u = 25t^2$ so that $F(t) = \int e^{-u} du = -e^{-25t^2}\big|_0^t = 1 - e^{-25t^2}$ so that $R_i(t) = 1 - (1 - e^{-25t^2}) = e^{-25t^2}$.
Then $R_s(t) = (e^{-25t^2})^3 = e^{-75t^2}$.

14.11. What is the failure rate for the time to failure distribution given in Exercise 14.10, if it can be obtained? If it cannot be obtained, explain why that is the case.

Solution:
The hazard function cannot be obtained because an expression for $F(t)$ cannot be obtained for the normal distribution.

14.13. Obtain the expression for the failure rate for a normal distribution if it is possible to do so. If it is not possible, explain why not.

Solution:
It is not possible to obtain an explicit expression for $F(t)$ because it is not possible to integrate the *pdf* of a normal distribution. It is possible, however, to obtain *some* expression for the failure rate. To wit, let $\Phi(\frac{t-\mu}{\sigma})$ denote the *cdf* for the time to failure for a normal distribution. Then $R(t) = 1 - \Phi(\frac{t-\mu}{\sigma})$. It follows that $h(t) = f(t)/R(t) = f(t)/(1 - \Phi(\frac{t-\mu}{\sigma}))$ with $f(t)$ denoting the normal *pdf*. We can simplify this by letting $\varphi(\frac{t-\mu}{\sigma})$ denote the standard normal *pdf*, so that we obtain
$$h(t) = \frac{(1/\sigma)\varphi(\frac{t-\mu}{\sigma})}{1 - \varphi(\frac{t-\mu}{\sigma})}$$
with $1/\sigma$ in the numerator resulting from the fact that σ is in the normal *pdf*, but not in the standard normal pdf.

14.15. Show that the failure rate for the extreme value distribution given in Section 14.5.5 is not a constant.

Solution:

$$h(t) = f(t)/R(t) = \frac{\frac{1}{b}exp[(t-a)/b]\,exp[-exp[(t-a)/b]]}{exp\,[-exp[(t-a)/b]]} = \frac{1}{b}exp[(t-a)/b],$$

which is clearly not a constant.

14.17. Consider the following statement: "For a series system comprised of independent components, the reliability of the system is less than or equal to the reliability of the lowest component reliability." Does that seem counterintuitive? Show why the statement is true.

Solution:
It is not counterintuitive because in a series system all of the components must function in order for the system to function. Since $R_s(t) = \prod_i R_i(t)$ and each $R_i(t) \le 1$, the reliability of the system cannot be greater than the lowest component reliability, with the two being equal only if all of the other $R_i(t) = 1$.

14.19. If possible, give three examples of engineering applications for which a constant failure rate seems appropriate, and thus the exponential model would seem appropriate as a model for time to failure. If you have difficulty doing so, state how the failure rates would deviate from being constant in engineering applications with which you are familiar.

Solution:
(student exercise)

14.21. A system need not be strictly a parallel or a series system, as many other combinations are possible and are in use. Assume that the Weibull model is being used and the system is such that three of the components are arranged as in a parallel system but the fourth component stands alone and must function for the system to function. Give the general expression for $R_s(t)$, assuming independence for the components.

Solution:
From the solution to Exercise 14.1, $F(t) = 1 - exp[-(t/\alpha)^\beta]$. With the components being independent,
$R_s(t) = (1 - (F(t))^3)(1 - F(t))$
$\qquad = (1 - \{1 - exp[-(t/\alpha)^\beta]\}^3)(exp[-(t/\alpha)^\beta])$,
which is the probability that at least one of the first three components will be functioning times the probability that the last component will be functioning.

14.23. Determine the *MTTF* for the lognormal distribution in terms of the parameters of the distribution.

Solution:

$MTTF = E(X) = exp(\mu + \sigma^2/2)$, with μ and σ^2 denoting the mean and variance, respectively, of $log(X)$.

14.25. Consider an exponential distribution with $\lambda = 2.4 \times 10^{-5}$ (hours) and determine the *MTTF*.

Solution:
The $MTTF = 1/\lambda = 1/(2.4 \times 10^{-5}) = 41{,}666.7$.

14.27. Determine the failure rate for a continuous uniform distribution defined on the interval (a, b). Does this seem to be a reasonable failure rate? Are there any restrictions on t, relative to a and b? Explain.

Solution:
$f(t) = \dfrac{1}{\beta - \alpha}$ $\alpha < t < \beta$. Then $F(t) = \dfrac{t - \alpha}{\beta - \alpha}$ so $R(t) = 1 - F(t) = \dfrac{\beta - t}{\beta - \alpha}$. The failure rate is then $\dfrac{f(t)}{R(t)} = \dfrac{1}{\beta - t}$. This failure rate is a strictly increasing function of t, subject to the restriction that $t < \beta$. Such a failure rate could be useful.

14.29. Consider Example 14.1 on the LEDs. Do you think a normal distribution would generally be a viable model for such applications? Why or why not?

Solution:
Since LEDs are new, it might be awhile before there is enough data available to suggest a failure distribution for them. A normal distribution may or may not provide a satisfactory approximation to the true distribution.

14.31. A safety alarm system is to be built of four identical, independent components, with the reliability of each component being p. Two systems are under consideration: (1) a 3-out-of-4 system, meaning that at least 3 of the components must be functioning properly, and (2) a 2-out-of-4 system. Determine the system reliability for each as a function of p. If the cost of the second system is considerably less than the cost of the first system, would you recommend adoption of the second system if $p \geq .9$? Explain.

Solution:
The reliability of the 3-out-of-4 system is $\binom{4}{3}p^3(1 - p) + \binom{4}{4}p^4 = p^3(4 - 3p)$. The reliability of the 2-out-of-4 system is $\binom{4}{2}p^2(1 - p)^2 + p^3(4 - 3p) = 6p^2(1 - 2p + p^2) + p^3(4 - 3p)$. Assume that $p = .9$. The second system has a reliability of .9963 compared to the first system, whose reliability is .9477. This improvement comes at a lower cost, so the second system should be adopted.

14.33. The first ten failure times for a repairable system are (in hours), 12150, 13660, 14100, 11500, 11750, 14300, 13425, 13875, 12950, and 13300. What is the estimate of the *MTBF* for that system?

Solution:

The *MTBF* is the average of the 10 failure times, which is 13,101.

15

Analysis of Categorical Data

15.1. The following survey data have been listed by Western Mine Engineering, Inc. at their website:

	Number of Mines	
	(Union)	(Non-Union)
Wages increased in past 12 months	58	70
Wages decreased in past 12 months	0	3
No change in wages in past 12 months	18	44

Can the hypothesis that wage changes are independent of whether a mine is a union or a non-union mine be tested with a χ^2 test? If not, can the table be modified in a logical way so that the test can be performed? If so, perform the modification and carry out the result. What do you conclude if you performed the test?

Solution:
A χ^2 test cannot be used with the data as given because 2 of the 6 expected frequencies would be close to zero. The most appropriate way to collapse the table is obviously to combine the second and third rows, so that the new second row becomes "no change or decrease", as shown below.

	Union	Non-Union
Wage Increase	58	70
No change or decrease	18	47

By hand computation using Eq. (15.2) or with appropriate software we obtain χ^2 = 5.607, and the corresponding p-value is .018. Thus, the evidence suggests that the two classification variables are not independent, which is also suggested by the fact that the two ratios of the numbers in the first row to the corresponding numbers in the second row differ considerably.

15.3. On February 12, 1999 the United States Senate voted on whether to remove President Clinton from office. The results on the charge of perjury are given in the following table:

	Democrat	Republican	All
Not guilty	45	10	55
Guilty	0	45	45
All	45	55	100

(a) Assume that someone decided (naively) to do a chi-square test on these data to test whether party affiliation and decision are independent. From a purely statistical standpoint, would it make any sense to perform such a test since the U.S. Senate was not sampled, and indeed this is the vote tabulation for all senators? Critique the following statement: "There is no point in performing a chi-square test for these data since we would be testing the hypothesis that Democrats and Republicans vote the same and I can see from inspection of the table that this is not true."

(b) Assume that the following counter argument is made. "No, this *is* a sample from a population since this is not the first time that the Senate has voted on whether or not to remove a president, and it probably won't be the last time. Therefore, we can indeed view these data as having come from a population --- for the past, present, and future." Do you agree with this position?

Solution:

(a) It is true that this is *not* a sample of count data, so a conclusion may be drawn simply from the table. No inference is necessary (or appropriate) because this is not a sample.

(b) Such a position is rather shaky such since such votes of impeachment are quite rare. Furthermore, this is not a sample from an *existing* population.

15.5. Give an example of a 3 x 3 contingency table for which the value of χ^2 is zero.

Solution:
One example would be

12	16	18
24	32	36
48	64	72

as the second and third rows are

multiples of the first row.

15.7. A company believes it has corrected a problem with one of its manufacturing processes and a team that worked to correct the problem decides to collect some data to test the belief. Specifically, the team decides to compare the results for the week before the correction with data they will collect during the week after the correction. The team collects that data and the results are summarized in the following table.

	Conforming	Nonconforming
Before	6217	300
After	6714	12

(a) Could a contingency table approach be used to analyze the data? Why or why not?

(b) If so, will the results agree with a test of the hypothesis of two equal proportions, using methodology given in Chapter 6? Explain.

(c) Perform a contingency table analysis, if feasible.

(d) Would it be possible to use Fisher's exact test? Why or why not?

Solution:

(a) Yes, a contingency table approach could be used as we may view the set of data as having come from a population and then classified into each of the four cells.

(b) Yes, the results will agree, as can be shown by working the problem both ways and recalling from Section 3.4.5.1 that the square of a standard normal random variable is a chi-square random variable with one degree of freedom. This can be explained as follows. If approached as a test of two proportions, the null hypothesis would be $H_0: p_{before} = p_{after}$. The test statistic was given in Eq. (6.4) as $Z = \dfrac{\hat{p}_1 - \hat{p}_2 - 0}{\sqrt{\hat{p}(1-\hat{p})(\frac{1}{n_1}+\frac{1}{n_2})}}$. We obtain $Z = 16.7845$ for this problem, the square of which is 281.72. This is the same value of χ^2 that results when the data are analyzed as a 2×2 table. The proof of this equivalence is lengthy and will not be given here.

(c) We obtain $\chi^2 = 281.72$ with a p-value of .000. Thus, we conclude that the percentage of nonconforming units before the corrective action differs from the percentage after the corrective action. (Of course this conclusion could have been drawn just from inspection of the table by looking at ratios of corresponding row elements.)

(d) Fisher's exact test could not be used for this example because the row and column totals are not fixed.

15.9. Assume a 2×2 contingency table and state the null hypothesis in terms of proportions.

Solution:
This may be stated in one of various ways, including stating that the proportion of counts for the first column that is in the first row is the same proportion as for the second column.

15.11. Generate 98 observations from the exponential distribution (see Section 3.4.5.2) with $\beta = 10$. Test the hypothesis that the data are from a population with a $N(10, 2)$ distribution. Use the power-of-2 rule to determine the number of equi-probability classes. What do you conclude? Would you anticipate that the result would be sensitive to the number of classes used? Explain.

Solution:
(simulation exercise to be performed by reader)

15.13. Fill in the missing value in the following 2×3 contingency table such that the value of χ^2 is zero.

20	35	18
24	42	

Solution:
The missing value should be 21.6.

15.15. Consider a 2×2 contingency table and explain why |observed − expected| is the same for each cell.

Solution:
The expected frequencies in each row and column must add to the sum of the observed frequencies for each row and column. This means that the differences between observed and expected must add to zero for each row and column. Therefore, if the difference between the observed and expected in the first cell (i.e., first row and first column) is, say, k, then the other difference in the first column must be $-k$, as must the other difference in the first row. It then follows that the difference in the second row and column must be k. Thus, all of the differences are the same except for the sign.

15.17. The faculty of a small midwestern university is surveyed to obtain their opinion of a major university restructuring proposal. The results are summarized in the following table:.

	Support	Opposed	Undecided
Men	116	62	25
Women	84	78	37

(a) Assume that every faculty member was surveyed. Would you perform a contingency table analysis of these data? Why or why not? If the analysis would be practical, perform the analysis and draw a conclusion.

(b) Now assume that a random sample of the faculty was used and the results in the table are for that sample. Now would you perform a contingency table analysis? Why or why not? If the analysis would be practical, perform the analysis and draw a conclusion.

Solution:

(a) A contingency table analysis should not be performed because the data are for a population.

(b) $\chi^2 = 9.32$ and the p-value is .01. Thus, there is evidence that the views of men and women differ.

15.19. Consider a 2 x 2 contingency table and the difference in each cell between the observed and the expected value. What is the sum of those differences over the four cells? What is the sum of the absolute values of those differences if one of the differences is 8?

Solution:

The sum of the differences is zero; the sum of the absolute value of the differences is 4(8) = 32.

15.21. Agresti (2002) (see References) gave a 2 x 2 table in which the entries are proportions that add to 1.0 across the rows. Is this a contingency table? Why or why not?

Solution:

This is not a contingency table because a contingency table can only contain counts.

15.23. Assume that a 2 x 2 table is included in a report with the cell proportions given. The report sustains water damage, however, and only the row totals and one of the cell values is discernible. Assume that the proportions are proportions relative to fixed column totals of 1.0, with the row totals not summing to 1.0. Can the numerical value of the χ^2 statistic be computed? Why or why not?

Solution:

Yes, the other three proportions can be obtained as follows. For the cell proportion that is visible, the other cell entry in that row can be obtained by subtraction since the row totals are known. The two missing column values can then be obtained by subtraction since then the column totals are 1. The analysis can then be performed on the proportions and the results will be the same as if the counts have been used.

15.25. Consider the following 2 x 3 contingency table for arbitrary factors A and B:

$$A$$

		1	2	3
B	1	30	58	67
	2	24	50	55

Without doing any calculations, explain why the use of Eq. (15.2) will not result in the rejection of the null hypothesis that the factors are independent.

Solution:
We can see from inspection that the ratio of the numbers in the first row to the corresponding numbers in the second row is almost constant. This will result in a small chi-square value and the null hypothesis will not be rejected.

15.27. The following 2 x 2 contingency table was given by Agresti (2002) (see References), with the source of the data being the Florida Department of Highway Safety and Motor Vehicles and the data from accident records in 1988.

Safety Equipment in Use	Injury Fatal	Nonfatal
None	1,601	162,527
Seat belt	510	412,368

(a) What is the hypothesis that would be tested?

(b) Would Fisher's exact test be applicable here? Why or why not?

(c) Perform the appropriate test and draw a conclusion.

Solution:
(a) The null hypothesis is that there is no connection between seat belt usage and the proportion of accidents that are fatal.

(b) Fisher's exact test would not be applicable because neither the row totals nor the column totals are fixed.

(c) A chi-square test can be performed and the results are as follows. As expected, the test shows a relationship between fatal accidents and seat belt usage.

```
Expected  counts  are  printed  below  observed  counts.  Chi-
Square contributions are printed below expected counts.

              Fatal    Not-fatal    Total
      None    1601      162527     164128
              600.51   163527.49
              1666.880    6.121
```

```
Seat Belt      510      412328      412838
             1510.49   411327.51
              662.685   2.434

  Total       2111      574855   576966

Chi-Sq = 2338.120, DF = 1, P-Value = 0.000
```

16

Distribution-Free Procedures

16.1. Consider the sign test in Section 16.2.2 and the example that was given. Explain why the test could not be used to test the mean of a population if nothing were known about the distribution for the population.

Solution:
If a sign test were used, there would be the tacit assumption of a symmetric distribution, so that the probability of an observation being below the mean is the same as the probability of an observation falling above the mean. Such a test could not be implemented without some knowledge of the population distribution.

16.3. Show that the reciprocal of the variance of the ranks is the constant in Eq. (16.2).

Solution:
We want to show that $\sum_{ij}(R_{ij} - \frac{N+1}{2})^2/(N-1) = N(N+1)/12$.

$$\sum_{ij}(R_{ij} - \frac{N+1}{2})^2 = \sum_{ij}R_{ij}^2 - (N+1)\sum_{ij}R_{ij} + \frac{N(N+1)^2}{4}$$

$$= \frac{N(N+1)(2N+1)}{6} - \frac{(N+1)N(N+1)}{2} + \frac{N(N+1)^2}{4}$$

$$= \frac{N(N+1)}{12}(N-1)$$

so $\sum_{ij}(R_{ij} - \frac{N+1}{2})^2/(N-1) = \frac{N(N+1)}{12}$, as was to be shown.

16.5. Describe, in general, the condition(s) under which the Friedman test should be used instead of the Kruskal-Wallis test.

Solution:
The Friedman test is the nonparametric counterpart to two-way Analysis of Variance and is thus used when two factors are involved in an experiment, whereas the Kruskal-Wallis test is the counterpart to one-way Analysis of Variance and is used when there is only a single factor.

16.7. The data in CLINTON.DAT contain a conservative score/rating that had been assigned to each U.S. Senator at the time of the Senate votes on perjury and obstruction of justice alleged against then President Clinton. Assume that you want to test the hypothesis that the scores for the Democrats and the Republicans have the same distribution.

(a) Test this hypothesis using the appropriate test (as if we didn't know what the conclusion must be!).

(b) Having done this, you have concluded, subject to a certain probability of having reached the wrong conclusion, that the distributions either are or are not different. But don't you know whether the distributions are the same or not in terms of location just by inspecting the data? Can the data for each party be viewed as sample data from some population? Why or why not?

Solution:
(a) Normal probability plots for Democrats and Republican scores exhibit non-normality, so the Mann-Whitney test is used. The results, show that the distributions differ.

```
Mann-Whitney Test and CI: Republicans, Democrats


Republicans N =   55  Median =   80.000
Democrats    N =   45  Median =    8.000
Point estimate for ETA1-ETA2 is 72.000
95.0 Percent CI for ETA1-ETA2 is (64.004,76.003)
W = 4006.0. Test of ETA1 = ETA2   vs   ETA1 not = ETA2
is significant at 0.0000. The test is significant at
0.0000 (adjusted for ties)
```

(b) Yes, it is obvious just from inspection that the scores differ greatly across the two parties. In order for the inference in part (a) to make any sense, we would have to regard the 100 senators as constituting a sample (admittedly non-random) from some population, such as recent sets of senators. Of course this would be somewhat of a stretch.

16.9. Assume that a sample of 60 nonsmokers and 40 smokers is obtained and whether or not each person has any type of lung disorder is recorded. Would you apply Fisher's exact test in analyzing the data? Why or why not? If you would not do so, what would you do instead?

Solution:
Fisher's exact test can be applied only if both the row and column totals are fixed, which is not the case here. Therefore, a regular χ^2 test could be performed or equivalently a test for the equality of two proportions.

16.11. Consider the data in 3WAYINT.TXT that was used in Example 12.3 in Section 12.5.2. Could that data be analyzed using one of the methods presented in this chapter? Explain.

Solution:
No, that was a 3-factor dataset and no methods given in the chapter can be applied to such data.

16.13. Consider the data in TABLE102.MTW. Use MINITAB or other software to construct a *loess* smooth. Do the data support a smooth for the entire range of X?

Solution:
The *loess* smooth is given below. The data do not support a smooth over its entire range because the data are quite sparse for $X < 60$.

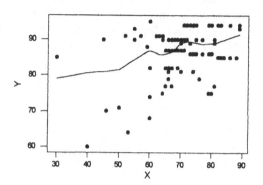

16.15. You are given the following bivariate data:

Y	23.5	23.8	24.2	25.1	26.2	27.8	26.1	27.2	24.2	25.1	24.2	23.3	24.6	27.0	25.2
X	17.9	18.3	16.9	17.1	19.9	20.1	18.4	18.2	17.5	17.3	17.0	18.9	18.4	19.2	19.1

Compute the value of the Spearman rank correlation coefficient.

Solution:
After ranking X and Y separately, the correlation between the ranks is found to be .494.

16.17. Consider a sequence of 14 data points with 10 points above the assumed mean for the population from which the sample was drawn. Does this suggest that the mean may not be the value that is assumed? Perform the appropriate test and draw a conclusion.

Solution:
The probability of at least 10 data points plotting above the assumed mean is .09 if the assumed mean is equal to the actual mean and the distribution is symmetric. This does not provide strong evidence against the assumed mean.

16.19. Like the independent sample *t*-test, the Mann-Whitney test can be applied wh
there are unequal sample sizes. One such example is given in *StatXact User Man*
Volume 1: a blood pressure study with 4 subjects in a treatment group and 11 subjects i
control group. The data are as follows.

Treatment 94 108 110 90
Control   80   94   85 90 90 90 108 94 78 105 88

With only this information, it would not be practical to use the independent sample
t-test as there is no way to test for normality when a sample contains only four
observations, and similarly a test of equality of variances would be impractical.
Therefore, apply the Mann-Whitney test and reach a conclusion.

Solution:
```
  Mann-Whitney Test and CI: X, Y

               N   Median
         X     4   101.00
         Y    11    90.00

Point estimate for ETA1-ETA2 is 9.50
95.7 Percent CI for ETA1-ETA2 is (-0.00,22.00)
 W = 45.0
Test of ETA1 = ETA2 vs ETA1 not = ETA2 is significant at
0.1027. The test is significant at 0.0981 (adjusted for
ties)
```

The test results show that there is no significant difference between the treatment
group and the control group.

17

Tying It All Together

17.1. An experimenter decides to test the equality of two population means by taking two small samples when the population standard deviations are unknown. He performs the *t*-test but later remembers that he forgot to check the distributional assumption. When he does so he sees that the two normal probability plots suggest considerable nonnormality. What would you recommend be done at that point?

Solution:
The Mann-Whitney test should be used.

17.3. Given the following graph, Would you fit a simple linear regression model to the data represented by these points, or would you use some other approach? Explain.

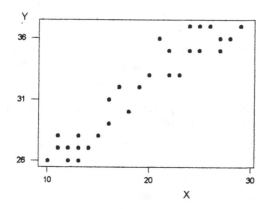

Solution:
The graph suggests that a linear relationship exists for only a relatively narrow range of *X*. Clearly a linear regression line should not be fit to the entire data set. One option would be to fit a line to only the subset of data that exhibits a linear relationship; another option would be to use a *loess* smooth.

17.5. What is the purpose in studying Bayes' rule in Chapter 3 and other probabilistic ideas and methods when these methods are not used directly in subsequent chapters?

Solution:
Probability forms the basis for statistical inference. Even though Bayes' Rule and other probability tools were not used directly in subsequent chapters, they are frequently needed in engineering disciplines and in other fields.

17.7. A small company suspects that one of its two plants is considerably more efficient than the other plant, especially in regard to nonconforming units. There is one particular product that has been especially troublesome for the first plant to produce, with the percentage of units that are nonconforming running at over 2% for that plant. The number of nonconforming units of the type in question is tabulated for a period of two weeks, with the following results.

Plant	Number Inspected	Number of Nonconforming Units
1	6420	130
2	7024	102

(a) A company employee will perform a test of the equality of the percentage of nonconforming units, which of course is the null hypothesis. What can be said about the inherent truth or falsity of the hypothesis for this scenario?

(b) There are three possible ways of performing the analysis. Name them. Which method would you suggest be employed?

(c) Use the method that you selected in part (b) and perform the analysis.

Solution:
(a) The null hypothesis is almost certainly false, as two percentages are not going to be equal if more than two decimal places are used.

(b) A test for the equality of two proportions (i.e., a Z-test) could be performed, or a test of proportions but without assuming approximate normality, or a contingency table analysis could be performed, after first constructing the table.

(c) The latter would be a reasonable choice as none of the counts are small. The analysis, given below, motivates a conclusion that the plants differ in terms of the percentage of nonconforming units,

```
       Chi-Square Test: Plant 1, Plant 2
  Expected counts are printed below observed counts

           Plant 1  Plant 2   Total
      1      130      102      232
            110.79   121.21

      2      6290     6922     13212
            6309.21  6902.79

  Total     6420      7024    13444

  Chi-Sq =  3.331 +  3.045 +
            0.058 +  0.053 = 6.488
        DF = 1, P-Value = 0.011
```

17.9. A process engineer plots points on a control chart and observes that six consecutive points plot above the chart midline, which is the target value for the process characteristic that is being plotted. She immediately concludes that the process must be out of control since the probability of six points plotting above the midline when the process is in control is $(1/2)^6 = 1/64 = .016$. What is the fallacy in that argument?

Solution:
The process may not be capable of meeting the target value. In general, neither the midline nor the control limits should be obtained from target values.

17.11. You are presented with data from a 2^3 experiment that was conducted to determine the factors that seemed to be significant and to estimate the effect of those factors. You notice that the AC and BC interactions are large, relatively speaking, in that each exceeds the main effect estimate for one of the factors that comprise the interaction. Explain how you will analyze the data.

Solution:
Conditional effects will have to be used (estimated), for at least some of the factors. How the data are partitioned for the conditional effects will depend upon the relative magnitude of the two interactions if conditional effects for C must be computed, and will also depend on for which factors the conditional effects must be computed.

17.13. A practitioner has data on six variables that are believed to be related, with one clearly being "dependent", in a regression context, on the other five. All of the pairwise correlations for the six variables are between .50 and .92, and the dependent variable is binary. The practitioner develops a multiple linear regression model using all five of the available predictor variables, then constructs a confidence interval for the parameter associated with the third variable and makes a decision based on the endpoints of that interval. Two mistakes were made: what were they?

Solution:
All five available regressors should almost certainly not be used because of the intercorrelations. Because the regression coefficients are probably uninterpretable because of the intercorrelations, confidence intervals based on these estimates are also likely uninterpretable.

17.15. An experimenter favors nonparametric methods, claiming that they are much easier to use than parametric methods because she doesn't have to worry about any assumptions, and thus doesn't have to spend time checking assumptions. Do you agree that assumptions can be dispensed with when nonparametric methods are used?

Solution:
No, nonparametric methods are also based on certain assumptions, such as the observations having come from a symmetric distribution. Since assumptions cannot be dispensed with entirely when nonparametric methods are used, a user might as well test the assumptions for the corresponding parametric test, and use the latter, which is more powerful, when the parametric assumptions appear to be at least approximately met.

17.17. The Dean of Students at a small university is interested in determining if fraternity or sorority membership is having an adverse effect on grades. He is interested in looking at grade point averages of at least 3.0 and below 3.0, with the latter in his mind being the value that separates good from not-so-good performance. The university's data recording system and computer systems made it easy to compile the following numbers, which are for the entire university, for one semester.

Status	≥ 3.0	< 3.0
Member	564	838
Nonmember	1239	1473

(a) Is this a sample from some population? If so, what is the population?

(b) *If* this is a sample from some population, perform the appropriate analysis and draw a conclusion, justifying the analysis that you performed.

(c) What analysis would you have recommended if the grade point averages had not been dichotomized?

Solution:

(a) It is reasonable to think of the data from one semester to be a sample, although not necessarily a random sample, from recent semesters.

(b) The analysis of the 2 x 2 contingency table is shown below.

```
             Chi-Square Test: C1, C2

    Expected counts are printed below observed counts

                    C1        C2      Total
             1      564       838      1402
                    614.44    787.56

             2      1239      1473     2712
                    1188.56   1523.44

        Total       1803      2311     4114

        Chi-Sq =   4.141 +   3.230 +
                   2.141 +   1.670 = 11.182

           DF = 1, P-Value = 0.001
```

We thus conclude that fraternity and/or sorority membership does affect grades.

(c) A test of the equality of the two population means could have been performed if the data had not been dichotomized.